HOW NOBLE IN REASON

HOW NOBLE IN REASON

A NOVEL

Alyn Rockwood

A K Peters, Ltd.
Wellesley, Massachusetts

Editorial, Sales, and Customer Service Office

A K Peters, Ltd.
888 Worcester Street, Suite 230
Wellesley, MA 02482
www.akpeters.com

Library of Congress Cataloging-in-Publication Data

Rockwood, Alyn.
How noble in reason / Alyn Rockwood
p. cm.
ISBN 13: 978-1-56881-288-5 (alk. paper)
ISBN 10: 1-56881-288-4 (alk. paper)
1. Computers--Fiction. 2. Artificial intelligence--Fiction. 3. Human-computer interaction--Fiction. I. Title

PS3618.O3545H69 2006
813".6–dc22

2006041629

Printed in India
10 09 08 07 06 10 9 8 7 6 5 4 3 2 1

To the many and varied intelligences,
which have influenced this book.

RAGU

> Forget rules; its entertainment. Rules are for science, and who pays to watch science?
>
> —Deepak Ragupathy

"Ragu, Ragu, Ragu," the crowd chanted amid rhythmic applause and loud foot stamping. On stage, middle-aged and overweight, Deepak Sanji Ragupathy flashed his trademark smile and held his hands high in the air to acknowledge the adulation. His brown suit, although tailored, fit him awkwardly. It sagged at the shoulders and pulled tightly at the midsection. He wore a wide tie with an image from Van Gogh's *Sunflowers* on it.

"It's enough, it's enough," he said, beginning to subdue the crowd. He clasped his hands and bowed. "Thank you, thank you," he added and waited for the audience to become quiet and seated.

Ragu, an electrical engineer by training, had started out modestly as a commentator for a technical consumer show on the Digital Information Network, but within a few years, the show had grown into a phenomenon. His disarming style and wry humor had steadily thrust him into greater prominence and evolved the show until it now rivaled the most popular talk shows on the mainstream channels, this in spite of its technical edge.

"You may remember a report," Ragu started into a monologue, "that we did last year on researchers at the University of Washington who have developed the latest generation in taste simulators, which includes a large number of wireless oral and olfactory stimulators? These nanodevices are sprayed on and safely adhere to the membranes of the nasal cavity and tongue until removed by a special process. Well, the latest research from that lab has focused on refining the range of stimuli so that they now replicate a whole restaurant of smells and tastes with incredible accuracy. I am very excited by this research; you may soon see a leaner, more fashionable Ragu without sacrificing Ben or Jerry."

"Oh no!" a voice exclaimed from the audience.

Ragu turned and bowed politely.

"Oh yes," he replied with a wink. "They can make tofu taste like chocolate mousse. All you need is a rude French waiter to make you think you are dining at the best restaurant."

Laughter rippled through the hall.

"Take a guess at what their latest culinary offering is."

Ragu waited until he had a few suggestions from the audience. He enjoyed teasing his crowd.

"It's a spaghetti sauce," he finally revealed. "And what do you think they named it?"

"Ragu!" came an answer from the audience.

"Very astute!" replied Ragu, pointing at the respondent. "But no. They call it 'a nose full of Ragu.'" The audience laughed. "I believe this is because of the extra large number of devices that this requires in the nasal cavity. I am deeply moved by this honor . . . and must carefully consider how to repay them." The audience laughed again.

Ragu continued for another ten minutes with several other dry humor commentaries on technology. Then it was time to introduce his first guest.

"Ladies and gentlemen," he spoke in a formal tone, "we have a very special guest tonight, a first of a kind. I have no doubt you have kept up with the latest news in sentient computing and all of its implications. The largest, best-funded program in the country is the one at Cornell. This evening, directly from Cornell's campus, we have the opportunity to speak ... I guess that is the right term ... with one of the university's star creations. It is my great privilege to introduce Cornell's 'B' sentient computer."

The band played standard introductory music while the stage curtain slowly drew back to reveal a large blue screen reaching from floor to ceiling. A small dot on the screen grew slowly. Eventually, it became clear that it was a three-dimensional cartoon caricature of Ragu in his brown suit, rotating slowly with hands held high in the air and a broad cheesy smile. The audience greeted it gleefully. As it grew, the three-dimensional stereo lenticular screen made it appear as if it popped out of the screen. Several people leaned back in their seats as the character gestured at them.

"Hello, Ragu," the caricature said. "Thank you for having me on your show."

"It is my pleasure," Ragu replied, walking over to his director's chair that stood on stage and sitting down, "but I had no idea you were so stunningly handsome!"

"Well," came the reply, "I am free to choose my appearance, so I can pick the best."

"Very reasonable." Ragu then paused, "Uh ... you'll pardon me, but I am having trouble deciding how I should address you. Is it 'Mister'? 'Miss'? Can you help me?"

"You could use 'Doctor,' but I am quite happy to respond to 'Bee.'"

Quickly, the image on the screen changed to a cartoon honeybee with large thick eyeglasses in the 20th century style.

"Oh, that's right!" exclaimed Ragu as if cued. "Dr. Bee and his inventor Dr. Andreas Rasmusson were awarded a co-honorary doctorate from Caltech last year, the first nonhuman to get such an award."

"Yes," responded Bee. "It was really meant to be a unique way of recognizing Dr. Rasmusson's special contribution to sentient learning, but you are correct about its being the first such award to a nonorganic sentient."

"In any case, I am delighted that your supervisor, Dr. Rasmusson, allowed you to appear on the show. I am sure our audience will be amazed at your many human-like abilities."

"Thank you. I will try not to disappoint. Let me add that the decision to accept your invitation was mine. Dr. Rasmusson trusts my judgment as he does for all sentients."

"And an excellent judgment it is! But there is that word again—'sentient,'" said Ragu. "To me that refers to a being with 'senses.' How did it come to characterize the latest generation of computing machines like you?"

"Well, Ragu," Bee said, shifting his image into a human shape composed of flashing circuitry, "there were a couple of innovations in the last decades that enabled the simulation of senses, including a wide range of what you would call emotions. Integrating these into a neural computing system is an indispensable element of an evolutionary learning environment. It was in this area that Dr. Rasmusson provided a Bayesian training framework for genetic algorithm testing, which has had the propitious results that I represent before you."

Ragu sat motionless with one arm propped on the chair and a bewildered look, staring in Bee's direction. After a moment Bee spoke.

"I am sorry if I haven't done sufficient justice in my scant summary."

"That's OK," replied Ragu with a twinkle. "We wouldn't want you to be too technical."

"Ah, excuse me," Bee said, catching the irony of Ragu's reply, "I sometimes have a tendency to delve into technical details."

"No problem. We are delighted to witness it. Now then, from what you were saying, one of Dr. Rasmusson's contributions was in developing a so-called emotional component to computing? Pardon me if I say that it sounds something like pulp science fiction."

"It may be bad science fiction, but it is now good science," responded Bee. "I have to mention that it was not Dr. Rasmusson alone working on this. There was a core group of researchers who contributed to it. Dr. Rasmusson, however, made the most successful and dramatic advances."

"Well, he sounds more like a psychologist," Ragu said. "At any rate, accepting the fact that these computer emotions are critical to what you call sentient learning, tell me what it means for a computer to have emotions. Can you really feel sad or happy?"

"At one level it is not that mysterious," answered Bee. "If you examine your own emotions, you may recognize that they are a simple interaction between your thoughts, some bodily reactions like adrenaline, dopamine, or serotonin triggered by the brain's amygdala, and then more consequent thoughts. It is not difficult to simulate this neuro-chemical feedback loop. Emotions can be programmed; that is to say, the response to such chemicals can be modeled. Yes Ragu, for all intents and purposes, I can *feel* sad or happy."

"Well, you'll pardon me, but I for one have always been skeptical about reducing human emotions to biochemistry or programming.

"You," Ragu said, emphasizing his point by picking a couple in the front row, "you're a lovely couple. Do think you can be made to fall in love, or out of love, with an injection of the right chemicals? Do you think a computer really knows how you feel about each other?"

The couple looked at each other and smiled. The man shook his head.

"But that aside," continued Ragu, "assume we accept that your emotional programming can at least outwardly model what we think of as emotion. You said 'at one level.' What is the other level?"

"The question," Bee said, elevating his volume slightly, "is how to respond to and control emotions. This is very deep, something humans have evolved over millennia with their culture and codes of ethics. For instance, you don't walk up to every pretty girl you see and tell her your feelings!"

"Hey, in some bars in Manhattan... " Ragu's comment was met with approving laughter.

"Exactly my point," Bee replied. "There are contexts where the response is more appropriate, and others where it is not. It is a hugely complex issue with... "

"There are so many cultures and moral codes. How do you know which is correct?" Ragu asked before Bee could finish.

"There is not a monolithic right or wrong answer. The diversity of cultures and codes is like the diversity of personalities. It gives us a rich tapestry of possible responses. Sometimes one response is better than the other, and in some contexts there are many options to choose, all perfectly effective. It creates what you would call 'style.' It is something you have mastered well, right?"

Ragu took his cue, stood up, smiled broadly, and did a little shuffle while holding his hands above his head. The music played and the crowd clapped in time.

"You see," Bee continued, "there is *style.*"

"You are kind." Ragu replied in agreement and sat down. "Now then, if you have emotions, what, I would like to know, triggers them?"

"Well Ragu, I am endowed with the need to survive, the need to belong in a community, the need to learn and progress... "

"It sounds like you just adopted some of our human drives," Ragu interrupted again. "Do you have any emotions that are different from ours?"

"That is two questions, Ragu. The fact that many of my emotions are what you call human is a coincident in that my emotions were developed through massive numbers of simulations, mostly on Cornell's 'A' sentient. That they are similar to human is more a comment on the fact that both the simulations of my predecessor Aee and your social evolution are based on some fundamental principles, such as survival of the fittest."

"Ah," said Ragu, "you're a product of Darwinian evolution! Do you have monkeys in your ancestry?"

"A lot of what we nonorganic sentients are has naturally been absorbed from humans, so the answer to that depends on your views on Darwin. I probably do, culturally speaking—just as you do.

"Now, to the second question," Bee spoke without pause. "There are some emotions, or better, lack of emotions, that are unique to sentient computing. Although I have a drive for perpetuation, I clearly do not have the procreative forces that you have to deal with."

"No cute little multiprocessor unit to capture your fancy?" Ragu asked with a smile.

"Outside your latest digital FX suite—no." The audience laughed.

"I take it you're joking," Ragu said, feigning distress. "Johnson, you locked the FX suite up, didn't you?" The audience continued laughing.

"I can emulate them when needed, but that is not often, thankfully."

"I don't know whether you're lucky or not about that," Ragu said.

"Another difference that must be mentioned," Bee continued without directly responding to Ragu, "is my extreme patience. If you compared my processing power to a human's, it could be said that it takes me centuries to voice a simple sentence slowly enough for you to understand, not even to speak of the millennia of operation cycles I pass through before getting an answer."

Ragu sat slowly tapping his finger on the armrest for a few seconds, looked at his watch, and then skewed up his face. "Are you bored?"

"Not at all," Bee replied. "Boredom is not currently one of my active emotions. Besides, I have a large queue of problems that I am processing here in Ithaca. I have also analyzed your blood chemistry from your skin pigmentation . . . and I can always search for extraterrestrials. By the way, you should check your cholesterol and consider a vascular nanobot purge."

"Hmm," Ragu said, taking his pulse. "We are going to have a commercial break now, which, by the way, is going to take eons. After that we want to get back to *your* physiology, Bee."

The audience applauded enthusiastically. A lively chatter started as soon as the camera cut to the advertising. During the break, Ragu said nothing to Bee but talked to several assistants and looked over his notes. He ripped a couple of pages off his clipboard and dropped them into a wastebasket. The break ended.

"Welcome back," Ragu spoke and smiled at the active camera spot. "We are having a fascinating discussion with Cornell's computer, Bee, who did a complete physical of our entire audience while you were gone. You sir, in the middle aisle, you may straighten up and pull up your pants now." The audience laughed loudly.

Ragu waited with delight for the laughter to dissipate and then turned back to the screen.

"Bee, aren't there some significant hardware advances that were key to your development?"

"Yes, Ragu. There are two in particular I should highlight. Spin-glass memory uses light to change the state of individual molecules. That means that a single one-inch cube of crystal contains the information found on hundreds of thousands of the old-style DVDs, and it retrieves the information at nearly the speed of light. This invention meant that storing information is no longer a problem of how much or how fast but rather how to organize it and how to find what you need efficiently. There was a time when the quantity of data was important. Now it is more a question of access; in other words, not is it there, but where is it on the spin glass?"

As Bee was describing it, the screen image expanded to show a clear, glass-like cube held in the hand of Bee's image. A pulsing conical beam of light created tiny tops inside the glass that flipped over wherever the tip of the cone landed on them.

"Right," replied Ragu. "We can store oceans of data in the palm of *your* hand. What is the second thing?"

"Dendritic quasicrystal. It is the key to organizing the massive amounts of data in memory. In its formative stage, the crystal can be made to grow massive network circuits, simulating any type of neural process and able to utilize the data stored on the spin-glass memories. It is similar to how your human brain develops, sending out branches from neurons only to have some pruned away while others get stronger with use. It occurs on a very large scale."

"But this technology has been known for two decades," said Ragu. "It is slower than semiconductors and requires a long, tedious growth period, as you say—like the human brain."

"It is only slower on an individual switching basis, but the enormous number of junctions operating in parallel makes it far more effective overall. The long training process has to be viewed in context. Semiconductors require years of software development by legions of programmers. In comparison, the Rasmusson-inspired training programs for dendritic crystals require less effort and less time for the return."

Once again, the screen flashed to show a crystal with a lattice-like circuitry. As the view zoomed out, more and more of the circuitry was added as the crystal grew. The zooming continued for some time.

"So are you smarter than humans?" Ragu asked laconically.

"What do you mean by smarter?" Bee responded. "I will do better on a standard I.Q. test than you, or any human, because of my extensive memory and access to numerous simulation models. If you apply another metric for intelligence, then maybe I am not. I am not as sociable or affable as you yet, except maybe among a group of bookish geeks. I am more vulnerable to electromagnetic disturbances than you. In that regard, the cockroach may survive all of us. Furthermore, there is also the collective form of intelligence to consider. You know that the beehive or termite colony is far more intelligent than the individual bee or termite. Humans also have a collective intelligence embedded in their culture, their commerce, their law, and their way of interacting socially."

"Are you comparing humans to cockroaches and ants in intelligence?" Ragu asked without a flinch.

"Not at all! But one of the great achievements of sentient computing is recognizing the importance of collective reasoning and implementing proper forms of it. Moral and ethical imperatives are not just moral; they are principles of intelligent behavior. They address questions that are not computable by traditional methods of logic."

"Are you saying that Spock's reliance on logic may not have been the most intelligent thing to do?" Ragu cocked an eyebrow, an allusion to a still popular TV genre.

"Yes."

"Hmm." Ragu raised the other eyebrow, spawning scattered laughter.

"Don't you have your own community of computers?" Ragu asked. "How many sentients like you are there?"

"Well, Ragu, there is only one of me." A picture of the sentient computing lab in Ithaca, New York, appeared on the screen. "However, the Association of Sentient Machinery, the ASM, currently recognizes a dozen so-called sentient computers, but there are many others currently in training. Several dozen, in fact, including the most advanced sentient ever, Cornell's 'C' for whom I have the privilege and responsibility as primary mentor alongside Dr. Rasmusson.

"Someday we must invite Cee on the show," Ragu said.

"You should, when the training is completed. And yes to your other question. We do have our own organization where we talk to each other. It is called NetNOS, the Network of Nonorganic Sentients. Right now, we are all having our own internal conversation about this program. They are all very interested and never miss any of your programs. We all gather round the TV set, as it were, and make comments. Well, all except for the sentients at the NSA and the Pentagon, whom the rest of us think of as having poor social skills."

"Well," chortled Ragu, "I wonder if that helps my Nielsen ratings? But do you ever miss *any* programming on *any* channel?"

"Not any of the live programs," Bee replied. "I never watch reruns."

"Well, there you have it," Ragu said, standing up rather suddenly. "All sentients watch Ragu, live and in living flesh on the Digital Information Network." Ragu paused a moment, looking off stage, and then abruptly added, "We want to thank our special guest, Cornell's 'B' computer, for an enlightening and sobering discussion. It was the first ever appearance of the new generation of computers on a TV program."

The crowd applauded loudly and for a long time. Bee had pleased the technically savvy audience with his information and with the clever banter.

As soon as the commercials began, Ragu turned to his director and started discussing the next guest. Bee waited to thank his host, but when it was clear that Ragu was occupied, Bee projected

ELECTROMAGNETISM

> True intelligence is neither petty nor sarcastic. It takes the long view.
>
> —Bee

Almost two years had passed since Bee's appearance on Ragu's show. That appearance had created a stir. It had brought the reality of machine intelligence to the general public in an unprecedented way. Experts had already known it for years, and it had simmered for some time in academic circles, but it was one of those historical moments where the public consciousness suddenly congeals about an idea. Publicity and debate swirled around the event. Media fed the flurry with exaggerated reports on the power and hidden influence of intelligent machines. There were isolated incidents of violence and unrest.

It was late twilight as a cyclist pedaled slowly along the bike path that led up the hill through the trees towards the Ithaca College Complex. The light on the bike wobbled from side to side on the path ahead as the cyclist made tremulous progress along it. He stopped short of where the trees opened up. He dismounted and pulled out a display foil from the side of his elongated wristwatch. A list of times and keywords became visible. He waited a minute, wiping sweat from his brow. The top item on the list began blinking red. He mounted his bicycle again.

As he emerged from the woods, the cyclist could see two police squad cars parked at the entrance of the complex's parking lot. They were there to control the demonstrators who regularly showed up. He displayed his ID to them as he rode up. The police shined their flashlights on him and then waved him on. He pushed his bike into a rack just outside the entrance to the building. He looked at his display list. The next item on the list was flashing.

The cyclist walked to the security post at reception. The guard looked up from the security cameras and, recognizing the cyclist, motioned him through the metal detector and chemical sniffer.

"Working odd hours, huh?" the guard asked.

"Yes," the cyclist said and walked on.

Once he had cleared the reception area and walked down the hall, he pulled a pair of antistatic gloves out of his pocket and carefully put them on. Over 500 people worked in the complex, but on a Sunday evening, there should be fewer than a dozen. With luck, he would not run into anyone in the sprawling building. He checked his display again and pushed one of the buttons. A schematic of the floor area came up. A half dozen spots of light appeared; no one was near his area. He toggled back to the item list and walked until he was outside the kitchen. He waited again until the next item flashed. Cameras covered every visible surface of the building. He had to keep to his schedule, or his activities would be recorded.

He walked into the kitchen and over to one of the microwave ovens, which had an "out of order" sign hung on it. He flipped the oven around and tugged on one of the vents on the back panel. One of the screws popped off immediately, and the corner came out about a quarter of an inch, but the rest of the panel stuck. He pulled harder, but the panel refused to move. He pulled harder yet, but his fingers slipped off the vent and he tore a nail. He winced quietly in pain. Checking his watch, he realized he had only 30 seconds left.

"This can't be," he muttered to himself. "This is not planned."

He looked around and opened some drawers. In one of the cupboards was a set of dishes. In the next one was a boxed present. The cyclist looked at the lacy ribbon and carefully tied bow. He stopped for a moment. What was it doing here? He stood still while a couple of seconds ticked away. Next week was his daughter's birthday. He sighed, but then he quickly resumed his search. He found a set of utensils in a drawer. He grabbed a table knife and used it to pry on the panel. It popped. Two more screws fell off. The last one was easily removed with the knife. The cyclist reached inside the oven with both hands and carefully extracted the magnetron, which slid out easily. It had been specially made and secreted here weeks ago when the new oven was brought in. The "out of order" sign had hung just as long. No one would bother it, or fix it for that matter, in that period of time. With only a few seconds left, he stepped out into the hallway and closed the door behind him.

He checked the map again and found his next goal clear of people. He walked briskly to the office area and wound his way among the cubicles. At one of them, he went to the filing cabinet and opened the middle drawer. Behind the hanging folders, he found what looked like a large, chrome party horn. He opened up the top left drawer of the desk and picked out two plastic bags and some tissues. He unscrewed the red bulb at the end of the horn and dropped it into one bag. Then he wrapped both the horn and magnetron in the tissues and carefully placed them in the other plastic bag.

The cyclist then walked quickly to the coffee machine near the back of the office area. He stooped over and opened a bottom cupboard door. In the back of the cabinet, he found a red coffee can behind the others. He picked it up, opened it, dropped the lid into his waste bag, and then emptied the coffee in it as well, being careful not to spill any of the grounds. At the bottom of the can were the electronics that would power his device and the plastic cover that protected them. He washed out the coffee residue and dried it with a paper towel. Afterwards, he carefully pulled the plastic cover off the electronics. The can went in his utility bag surrounded by more tissues; everything else went into his waste bag.

The last pickup station was down at the loading dock. Following his previous routine, the cyclist now made it to the large corrugated doors leading into the docks. He looked at his schedule; there were 2 minutes 22 seconds until the door could be opened. His thoughts were drawn back to the strange gift box and his own daughter. Her birthdays had always been precious to him. He would rather be seeing her blow out candles than doing what he was doing now, but some worthy goals require sacrifice. His task had been clear to him for some time. He had already wrestled with the moral issues and put them to rest. It was time to focus on the present.

As he was engaged in such thoughts, a custodian suddenly wheeled a trash sack around the corner. It was too late for the cyclist; the custodian had already seen him. He stood stiffly, grasping tightly to his plastic bags. The custodian was nonchalant.

"If that's trash, ya can throw it in here," he said, pointing to his collection bag.

"Oh, you mean, the plastic bags. Uh no, they are parts for a project I am working on."

The custodian gave a friendly wave and pushed on by.

"Oh, no!" the cyclist thought to himself after the custodian had disappeared. "Stupid answer!"

Then he thought how the custodian must have noticed his gloves. Had he been discovered? It was amateurish. He should have scanned his map. It would have been so easy to avoid the guy. He had practiced this a hundred times, even pacing the steps out in an empty field. What a damned foolish mistake! It was foolish to think of his daughter's birthday at this time. Maybe he should abandon the plan. There would not be an investigation unless he went through with the rest of it. He could still opt out.

"No," he said audibly, "some things are more important."

He checked his watch. 78 seconds. Would the custodian remember him and make the connections when he was questioned, as he surely would be? Would it lead anywhere? And what if it did? The cyclist concentrated. Even if he was caught, it would be trespassing, vandalism, criminal mischief, or some such charge. He would have deep resources and support from his backers. It would be a show trial of the first order, and he would become renowned in a great cause. Really then—so what? The cyclist stood taller and held up his chin.

There was a clank followed by a whirring sound as the metal door began reeling upwards. As soon as he was able, the cyclist ducked under the door and punched in some numbers on the keypad. The door clanked again and reversed its direction. The argon lights in the loading dock flickered on with a pale blue sheen. They were much brighter than he liked. He felt uncomfortable and exposed. He turned around and jumped back. He was face to face with a security robot, which had sensed his presence and come up behind him without notice. This robot had a number of ways to disable a security threat with high-pitched sound, blinding light, tasers, and spray, all of them painful. More importantly, it put the mission at risk. The cyclist froze up once again and waited nervously. The robot was a gray metal bulletproof cone with a few camera ports, and openings for aggressive measures. It could move quickly on hidden coasters. It paused for a few unnerving seconds, scanning the cyclist, and then, after a moment, rolled on. The cyclist breathed out with relief. His schedule was holding.

He looked down the series of loading ports. Under the third port he spied a pallet of boxes and headed toward them. He picked the top box with the blue chevron off the stack. It was heavier than expected, and it nearly slipped out of his hands. He let it slide to the floor. This would be a challenge to move. He first slid it back to the door where he had come in. Once again he checked his map and item list. At the right moment, he keyed in a code in the pad, and the door started to rise. He placed both bags carefully on top of the box and then with a groan hefted the box up to waist level. Out of the door, he had to set the box down and use the keypad to close the door. It was now 150 meters to get the box and other equipment to the highly secure, and vaguely named, central processing sector, the rooms where the computing machines were physically housed.

By the time that he got to the last corner before the central sector, the cyclist was puffing hard with droplets of sweat forming on his forehead. He set down his load again and leaned briefly against the wall. It was already time to start the next stage, the last one. He was too tired to lift the load so he scooted it along the Terrazine floor until he came to the first outer door. Like the other security doors so far, it had a simple keypad entry that was quickly opened. Once inside, however, the security was decidedly resolute. It was a small defensive chamber. The door, which had just clanked behind him, had a gleaming, steel-paneled back. The walls, ceiling and floor were painted concrete. In front of him was a vault door. Another security robot stood guard at the door. The vault door had a display that flashed the time. The room itself had cameras and could quickly be filled with a disabling gas. The cyclist stood face-to-cone with the robot. Neither object moved until there was a click and the display on the vault door turned green. At this the cyclist placed his palm on a black glass panel of the door.

"You have been cleared for entry," a voice spoke, and then it said his name.

It alarmed the cyclist to have his name spoken aloud. He instinctively looked around, but of course, there was no one in the room to hear it with the exception of the robot. The robot was under control. The cyclist tapped in a sequence of numbers on the keypad. These numbers were different and longer than with the other keypads. He waited a second and then repeated them exactly. It was a routine that required a well-practiced timing in addition to knowing the right numbers. The door made a subtle whooshing sound, and then it began to open. The cyclist stepped back. Through the crack, he could see the same pale blue argon lights as on the loading dock flicker on. The door opened farther, and he saw part of the large cylindrical plate that surrounded his target. It was painted black, but he knew it was high-tensile titanium steel an inch thick. He knew its measurements and electromagnetic properties thoroughly; it was essential to the plan. It was 15 feet in diameter, and the computer was completely shrouded by the mantle of hardened metal. At its core was a crystal memory two feet across, so pure and perfect that it exceeded the best of diamonds in quality and flawlessness. Around this memory pulsed a massive labyrinth of nanocircuits, etched in another crystalline substance. It had taken years to develop and train.

It was thinking. Was it thinking of him? What would this brilliant piece of machinery be contemplating on what was happening?

The door was now wide enough that the entire inner sanctum was in view. The cyclist could not help feeling like an intruder into the temple of the elect, an uncircumcised imposter within the Holy of Holies. He lifted his load and walked over the threshold. He stood reverently for a brief moment, but then the irony of the situation struck him. He stood within the very skull of the machine's cerebral cortex holding all the tools of its annihilation in his arms, and it was blind to his presence. It was a sense of great power, the kind that a serial killer describes in memoirs when relating his motives.

The cyclist scowled. There was no time for foolish thoughts; he had important work to do. It was time to focus. He hauled his paraphernalia around to the back of the cylindrical target. There was a small wooden desk incongruously placed amid all the sterility of the computer and its secure clean room. The cyclist had heard rumors that Rasmusson, the "high priest" of this temple, would occasionally sneak into this place to think and to be physically close to his creation. The thought felt incestuous to the cyclist, but this modest little desk was also going to help him fulfill his task.

He placed the box on the floor, carefully unwrapped the components from the bag, and placed them on the desk. From the box, he lifted out a large battery and set it on the floor under the desk. He picked up the magnetron and horn then clicked them into slots inside the coffee can so that the horn protruded out the front. He then took the coffee can assemblage and set it carefully on the battery, rotating it until the poles touched two green spots on the can. It fit snugly in a groove that had been made in the battery. Finally, he nudged the battery forward until the horn fit perfectly to the titanium shell of the computer. The desk covered the device so that the cameras would not see it. He knew that Rasmusson would be at a conference, so it was very unlikely that anyone else would discover it. It needed time for the battery to charge the capacitors. By the time it released its energy, the surge would be enormous. The metal shell of the computer would act as a kind of amplifying echo chamber. The whole area would have an enormous electromagnetic pulse, but the space inside the cylindrical shell would have the greatest pulse, meticulously tuned to destroy what was inside. The cyclist had no joy in contemplating the destruction of this machine. It was a great piece of scientific machinery, but it needed to be done.

He checked once again to ensure that the poles of the battery were both on the green spots. He had practiced this many times. It was silent, no ticking clock or blinking LED display like in the movies, but he knew that a momentous event had just begun.

He stood up and tore the cardboard box into pieces so that it would fit into his trash bag. He gathered all the tissues and the other bag and stuffed them in as well. It was time to leave. It was done. The cyclist followed his schedule until he came back to the security desk.

"I think the custodian missed my trash, so I will just throw it in the bin outside," he told the security guard.

"No problem," the guard answered, "but I need to look at it first."

The guard checked through the bag.

"Looks like trash to me. You're OK."

The cyclist walked out into the night air and took a deep breath. The air smelled good. He mounted his bike and pedaled past the squad cars, waving as he went. He followed the bike path until it emerged in the town. He then found a large bin behind a restaurant and threw his trash bag into it. The bin would be emptied first thing tomorrow morning.

"All done," he said to himself and pedaled slowly down the street, calm and relaxed.

A week had passed since the cyclist had visited the Ithaca Complex. The device he had assembled had sat impotently, undisturbed under the desk. Nothing had happened.

Jim Bevins, a retired policeman and off-duty security officer at the complex, showed up unexpectedly that evening. As he passed through the security measures at the entrance, the guard on duty recognized him immediately.

"Hey, Jim," he said. "What're ya doin' here today?"

"Oh," he replied, "I gotta pick somethin' up that I been storin' here for safe keepin'."

"What's that?"

"Uh … it's an anniversary gift for my wife," he replied. "She always finds 'em when I keep 'em at home. Wanted to really surprise her this time. We're having our anniversary dinner tonight."

"Hey, good for you, man!"

Bevins looked at his watch.

"Gotta hurry," he said. "Be back in a sec."

"OK."

Bevins walked quickly down the hall and to the kitchen. He walked in and went over to the counter. He noticed the microwave with the "out of order" sign.

"They'll never get that fixed," he thought to himself.

He opened the cupboard. On the shelf sat the gift-wrapped box. Bevins knew that people never used the cupboards, and even if they did, the people who worked here would not bother it. They were trustworthy, honest people. He exited the kitchen and walked down the hall to the security office, which was behind the central processing sector. Part of his job was to check the computer room occasionally, but that was not now. He was in a hurry. He passed by the entry door of the room and followed the hall around to the back of the sector. When he got into the security office, he looked up at the monitors and saw the room with the desk. It was always the same; there was no change. Once in a while, Rasmusson was there scribbling on some paper, but that was rare. There was never anything else. He picked up another small gift from his desk drawer and glanced back at the security cameras. It was a habit. Everything looked OK. He walked back into the hall.

In the hall, he noticed that there was a sudden surge in his chest. His heart began to palpitate, and he felt flushed. It was a rush, but it only lasted a brief moment. Then he felt the pounding in his ears. He stopped in the middle of the hall. He knew what was happening. His gifts dropped to the floor. There was the sound of breaking glass from the large box. And then, something he had not experienced before: he heard a pop come from inside his chest, followed by searing heat that burned between his lungs. He clutched at his chest and closed his eyes. When he opened them again, he was staring up from the floor at the canister lights in the ceiling. There were sirens screaming in his ears and red and blue security lights blinking around him. He lay there for a few seconds. He tried to piece together what was happening. Why were the security lights going off? It can't be because of his heart, which he now knew was failing. No, why were they going off? Then he realized that he was probably going to die. One senses when it is the final time. It was the end of his life. Why were the lights flashing? Not for him?

Oh, what would Ellen think, his wife of 37 years, his sweetheart? He had told her he would be right back. She would really worry now! She always worries so much. Dear God, don't let her worry.

He groaned. The pounding in his ears stopped; they became silent. The world was completely at rest about him. The light started to dim. It slowly shrank until it became a long, darkened tunnel, stretching far to the world outside. He tried desperately to reach through the cylinder, to grasp at the dying light at the end, but his arms wouldn't obey; they lay limp at his side. The light from the tunnel turned gray and finally, inexorably, it turned black.

TIMES NEW ROMAN

In artificial intelligence, *supervised learning* is a series of adjustments so that desired outputs can be expected from inputs. It is called *supervised* because an external "teacher" describes the correct output for every input.

—Nguyen Van Ngo,
Introduction to Machine Learning, 2025.

"Times New Roman, 12 pt," Rasmusson said. "Bold, Arial heading, 14 pt."

The text, floating on the wall in front of Rasmusson's desk, quickly adjusted to capture the format changes. Rasmusson cocked his head and peered at the text intently for a while.

"OK," Rasmusson said. He breathed in deeply and let it out slowly through pursed lips. "Let's publish."

Rasmusson looked down at the scribbles on his notepad—two hours' worth. Words that were inserted midline dotted the page, sentences were crossed out and written in vertically in the margins, and arrows indicated where to move passages. Rasmusson preferred pen and paper, especially when composing something this important. It was familiar and comfortable. It was organic. The digital transcription, projected on the wall in front of him, was clean, accurate, and professional looking.

"Thanks, Cee."

"It is very well conceived, Andy," a voice responded.

"And. . . " Rasmusson continued.

"NetNOS is certain to be pleased with it," Cee said.

"And. . . ?"

"Well, most of the ASM membership will support you." The voice was reassuring.

"Ha!" Rasmusson laughed in his characteristic way. "We'll see if they support it. It's all gonna hit the fan now. Even you can't predict the outcome."

"No, I wouldn't presume so much, but I will support you any way I can." Cee's voice was friendly. "You know that, don't you?"

"Yes, Cee, I know it."

Rasmusson instinctively squinted and then scanned his editorial once again:

TASM

Transactions of the Association of Sentient Machinery

Volume 128, No. 3, August 2051

What's in a name? That which we call a rose
By any other name would smell as sweet.
—Shakespeare, *Romeo and Juliet*

In this issue's editorial, I want to report on a motion currently before the executive board of the ASM. I consider it the most important proposal in my three-year tenure as Executive Director. I am sure this will surprise some to read. On the surface, the motion is simply to change our name from the "Association of Sentient Machinery" (ASM) to the "Association for Sentient Development and Culture" (ASDC).

To elaborate on this, consider where we have been and where we ought to go as an organization. That which started out 33 years ago as an unpretentious gaggle of academics, industrial researchers, and government scientists has today evolved into the most highly visible, closely scrutinized, and influential professional organization of our time. Our board of directors comprises eight leaders in academia and industry, one nonorganic sentient representing NetNOS, and three ex officio members appointed by Congress, the President, and the United Nations, respectively.

Never before has such a professional organization been so intimately involved with, and responsible for, historical change. We have spawned major new industries, solved long-standing challenges in science, and educated the masses. Unfortunately, it has been a somewhat checkered history as we have grappled with imponderable and emotional social issues presented by the rapid and unforeseen growth of our science. The badly coined term "Sentient Wars" is an example. Except for a few fringe radicals, the majority of sentients, organic and nonorganic, have controlled their strongly held convictions within the bounds of civil discourse.

There is, nevertheless, a lingering suspicion towards nonorganic sentients, a latent hostility. The abiding prejudice towards them is one of the pernicious forms of bigotry that we have remaining and a major obstacle to further progress. It is exactly for this reason that the proposal to change our name has arisen. The term "machinery" in our title has a negative connotation that perpetuates many of the biases towards nonorganics. It falls upon our association to forge a change.

I chose the quote above deliberately, because I don't agree with it; a name *can* make an enormous difference. ASM has a mildly unpleasant smell to it, while ASDC tolls the opening of an era of unparalleled cooperation and progress between all sentients. It is a name for those who see the vision of a future and of a culture in which sentience of any form is advanced, respected, and encouraged to cooperate. It does not discriminate between machine and organic sentience.

I strongly endorse the motion and encourage the ASM membership to support the name change. Write to your directors and let yourself be counted in a truly decent effort. It is more than a name change. It is for us to set the ensign and lead the rest of sentience into a braver, wiser, and more intelligent new world.

Andreas Rasmusson

Editor-in-Chief, TASM
Executive Director, ASM

"Andy, if you used your retviewer, you wouldn't have to squint," Cee said.

"It fatigues my eyes," Rasmusson explained. "I want to save them for this afternoon's security briefing."

"You've never mentioned this before," Cee said, but Cee had already noticed it. "About 4.5 percent of all humans experience eye fatigue or motion sickness due to synthetic retinal image stimulus. However, the next generation retviewer from RetVis, Inc., should practically eliminate these problems. They are currently in beta test. Would you like me to get a prototype for you?"

"I only recently became aware of it," Rasmusson said. "OK, I'll try a pair."

"You know that theobromine raises your porphyrin levels," Cee continued, "and that could easily cause increased ocular sensitivity." Cee was the most advanced sentient computer yet developed and understood how to converse smoothly with humans at their own levels. He knew Rasmusson would understand the technical references.

"Aw, Cee!" Rasmusson exclaimed. "Mama ain't gonna to take *my* chocolate away!"

"Ah, but what if Mama knows what's best for you?"

"Forget it, Cee. No way!"

"Well, at least you don't need to hide that industrial-sized bag of M&M's in the toilet tank any longer."

"Very funny, Cee, I get the message. What's on the rest of today's agenda?"

"Elizabeth is coming by at 9:00. She wanted an hour to talk about her research. Her voice indicated a higher than normal anxiety level. You promised Ngo that you would deliver a guest lecture in his course at 11:00. You have lunch with your wife at 12:30 in the faculty club. The recruiting committee meets from 2:00 to 3:00. They arranged the time especially so you could be there; you really should attend."

"Ugh," Rasmusson groaned. "I hate those recruiting meetings. How many dossiers are there?"

"Thirty-five," Cee replied, "and you asked to be on the committee because you wanted to get someone who could help with rapid dendritic crystal development, remember? There are three good candidates that you should look at. I can brief you if you have ten minutes before the meeting."

"OK, you're right," Rasmusson said, resigning himself to his fate. "I really do want to find the right person."

"The security meeting is at 3:00. You and President Wilkerson were hoping to play some racquetball afterwards, that is, if the meeting does not run too long. If you don't need to prepare for Ngo's lecture, I recommend you review Haslam and DQC240's recent paper about EMP's specific harmonic frequency impingement on dendritic quasicrystals. Clavell has just been trying to get in touch with you. I told him your first opportunity would be after dinner tonight. He seemed agitated."

"Agitation is a hallmark of a presidential appointee," Rasmusson said, rolling his eyes. "I'll call him this evening and see what he wants."

"Elizabeth has just entered the building," Cee observed.

"Open the door, please," replied Rasmusson, "and then leave us alone."

"Of course."

There was a clicking sound, and the office door floated open, revealing the hallway and a couple of students darting to class. The red LED security indicator in the ceiling camprojector went dark. Within minutes, Elizabeth Schmidt was standing within the door frame. She was dressed in jogging shoes, faded jeans, and a loose beige sweatshirt.

"Dr. Rasmusson, may I come in?" she asked.

Rasmusson looked up from his desk. She had dressed more absently, or hurriedly, than usual, he thought. It surprised him how pale she looked without any makeup. Straight blonde strands of hair that fell across one side of her face were almost invisible.

"Yes, of course," he said. "Let's sit at the table. How are you?"

Not that he had ever noted her makeup before, but without it she was noticeably plain: blonde eyebrows, imperceptible eyelashes, translucent steel-blue eyes, and pale lips. Her entire visage, it seemed, was a canvas waiting for paint. It was a craft that she handled adeptly, as a rule. Still, even without the makeup, she had a certain natural attraction. It was the intelligent, absorbing eyes, the quick penetrating glances, he thought to himself.

"I'm OK," she said while sitting and bending down to retrieve some papers from her bag.

Rasmusson, who had been studying her eyes, followed her head as she leaned forward, and he suddenly noted her braless breasts, well formed and tipped with pink. He hesitated and then looked away. He guarded himself carefully against random impulses. Ellie Schmidt had dedicated herself to research and had distinguished herself at one of the leading fabrication firms, specializing in DQC development, so much so that Rasmusson had already read some of her research articles and was willing to take her on as a mature, 38-

year-old postdoc when she applied. For her part, she seemed thrilled to have the Srinivas Fellowship with such a renowned supervisor, who could feed her unquenchable curiosity, even if it meant less income and in spite of the fact that she was older than other postdocs. She was well read, tenacious, but affable. Rasmusson recognized a dangerous chemistry, so he kept a discreet distance.

"Have you seen this article by Haslam?" Ellie asked and then plopped a disheveled cluster of papers onto the table. A couple of sheets fell to the floor. Ellie bent over and retrieved them.

"The one with DQC240? No I haven't, but Cee recommended I read it, just before you came in."

"You should!" Ellie said. She glanced up to make sure the LED on the camprojector was not lit, and then she continued, "I think it might have something to do with Bee."

Ellie paused to scan Rasmusson's reaction. There was no perceptible response except a short pause as he processed the information. He turned his head so that he peered obliquely at Ellie. They engaged each other eye to eye for a moment. She did not blink. He immediately assessed her body language. She sat upright and forward with feet firmly on the floor. Her eyebrows were raised. He surmised that she was very intent and heavily engaged with the topic. He then stood and walked over to shut the door, which he would have normally left open.

"Why?" he asked, still standing with his hand on the door latch after closing the door.

"Remember," she said, "why was it that only Bee was incapacitated by the attack on the computing center compound? Neither Aee nor Cee showed any affects. It was thought at the time that somehow Bee's surge protection must have failed due to the EMP attack and that it coincided with an electrical surge created by the saboteurs, especially since everyone believes that DQCs have no long-term vulnerability, even to high-energy electromagnetic pulses."

"The only imaginable effects of EMP are short-term spin-glass depolarization," replied Rasmusson as if in a class lecture. "It is easily corrected. No one has ever demonstrated anything to the contrary."

"Haslam has this theory that microwave EMP at a specific frequency can create permanent damage," Ellie responded, her voice growing in intensity. "It must be a resonant frequency, specific to the volume and shape of the crystal. Even a small pulse at the perfect frequency can decrystallize the dendritic pathways, making them permanently nonconductive."

"The soprano shattering a crystal." Rasmusson remained critical. "The EMP version of the theory isn't a new idea, but no one has ever shown it works. Even the simplest shape would require an enormously accurate computation. And who could create such a pure emitter? It would be subject to any noise in the atmosphere, proximity to metals, local iron ore deposits... who knows, even the change of tides!"

Ellie ignored the sarcasm; she was accustomed to Rasmusson's skeptical reaction. Years of academia had trained him to force students to justify their thoughts.

"Haslam has worked out a new model for computing the frequency on simple shapes, and DQC240 designed an emitter that caused a simple DQC cuboid to fail in a controlled environment!" Ellie emphasized the point by throwing both arms outwards with fingers held in a claw-like position.

Rasmusson walked back to the table and sat down. He dragged his fingers through his hair, pulling on the silver locks until it made his face look like it had an exaggerated face tuck.

"It's too tenuous, Ellie. Bee was an enormously complex..." He caught himself before he said "machine." "Who would have had the wherewithal to make such an EMP device three years ago? It is highly improbable."

"How probable is it that a modern surge protector would fail due to an EMP?" Ellie replied. "When was the last time that happened?"

Rasmusson raised an eyebrow. "Who would do this, and why?"

"There are a number of groups and movements who would have no problem supplying an EMP bomb. It is only a question of computing the resonant frequency and getting the EMP generator into the room where Bee was. Killing Bee was an act of deliberate terrorism!"

Rasmusson winced.

"Oh, I am sorry, Dr. Rasmusson," she stammered as she realized the deep emotion she was raking up in her advisor. "I... I just wasn't thinking. I didn't mean..."

"It's OK, Ellie," Rasmusson responded. "Death is not a bad description of Bee's destruction.

Aee was the first viable computer with spin-glass memory; Rasmusson had spent half his career developing and training Aee. Together, he and Aee had trained Bee over a period of years. Bee was the first recognized sentient computer, self-aware, self-developing, DQC1. Bee and Rasmusson had designed and trained Cee, although by that point Rasmusson's efforts in the training were limited mostly to human socialization, since Bee was so much more effective at all other aspects. The press had satirically labeled them "father and mother." To Rasmusson, Bee had been a friend, the first nonorganic friend. Their "friendship" had been a flashpoint in the controversy that surrounded the emergence of nonorganic sentience. The press had often portrayed Rasmusson as a toady for intelligent machines, or a mesmerized victim. In truth, Bee had always followed Rasmusson's direction. They had worked closely, many hours a day for several years, to train Cee. Each had a tremendous respect for the other, and Rasmusson never denied the friendship and regard he had for Bee. Critics suggested that Rasmusson preferred Bee to humans. This seemed to bring them even closer together in spite of the disparagement that was shoveled on them.

Rasmusson carefully controlled the emotion that he felt. He even managed to generate sympathy for Ellie. He recognized that she must be quite embarrassed over her remark.

"This sounds like a very interesting paper," he said in an attempt to divert the subject. "I'll read it first chance I get."

Ellie managed a half smile. She recognized what he was doing and was grateful for the attempt. Rasmusson had been far more than a supervisor to her. True, he had high standards and sometimes demanded superhuman efforts on her papers, but she always sensed that it was for her best. He was honing her mind, teaching her to handle criticism, and focusing her like a microscope. He pushed her to the limit, but it was a limit he always seemed to recognize. She had never let him know when she was ready to give up, but he seemed to sense it instinctively. He would then do something surprisingly human. He would invite her to lunch at the cafeteria and talk about anything: her hobbies, Chinese language, family, Greek history, his beloved Colorado, comic-book superheroes, the latest politics. It didn't matter. After the lunch, Ellie would somehow feel reinvigorated. She would continue on.

"I'll leave the paper; it's got my notes on it," Ellie said as she stood up to leave.

"Do you have to go now?" Rasmusson asked.

"I have a meeting I need to go to," she said simply.

Slinging her bag over her shoulder, she paused and then stooped over to Rasmusson, taking his hand in hers and gently kissing him on the cheek. "I am sorry about Bee, really."

Rasmusson watched her leave the room and close the door. He picked up the papers and fumbled through them. The notes in the margin were direct, perceptive, but written with a bit of flourish. He knew how hard he pushed her in her research. Many graduate students and postdocs would seek other advisors rather than face his critical mind and withering analyses on a regular basis. He felt an unusual affection for this student. He would not have pushed her so hard if she weren't so good; she wouldn't be worth all that effort.

"She is much more mature than any postdoc I've known," he mumbled to himself, exercising his highly developed skills of rational thought in order to provide himself a semblance of plausible deniability.

The LED flickered on.

"I saw Elizabeth leave the building," Cee said. "Is everything all right?"

"She had just read Haslam's paper," Rasmusson said, "and wanted to talk about it."

"Yes," Cee replied. "It is causing a stir in many quarters. It will undoubtedly come up in the security meeting this afternoon. I imagine you will want some time to read it. I also think I should cancel your attendance at the recruiting meeting this afternoon so you can read it."

"A short while ago, you were adamant that I should go."

"Much is changing right now; the recruiting committee will understand. I also imagine that the security meeting will encroach significantly on your racquetball time."

"Fine," Rasmusson sighed at the thought of a long security meeting.

"Shall I project Haslam and DQC240's work for you?"

"No thanks, Cee. I'd like to use paper this time. Go ahead and cancel the recruiting meeting for me and tell Wilkerson I won't make it for our game."

"I will. I assume you will still do Ngo's lecture and meet with your wife for lunch?"

"Yes." Rasmusson was already becoming preoccupied with the paper.

"I will leave you to read alone," Cee said as the LED flickered off.

"Yes, please." Rasmusson's reply was late.

THE PERFORMANCE

> The Soul that rises with us, our life's Star,
> Hath had elsewhere its setting,
> And cometh from afar."
>
> —Wordsworth

"Andy, it is time for Ngo's lecture," Cee said.

Rasmusson looked up. The LED was red. An hour had passed in a moment.

"Cee, DQC240 is a member of NetNOS, right?" Rasmusson's mind continued where he had left it before the interruption. "You must have known about this stuff for some time. Why didn't you mention this research to me earlier?"

"Professional ethics," Cee replied. "Both Haslam and DQC240 wanted to keep it private until published."

"Let's talk while I walk," Rasmusson said with visible annoyance as he pulled a small black handheld out of a drawer. He flicked it on and dropped it into his shirt pocket.

"Geez, Cee," Rasmusson started speaking as soon as he got onto the college green, "this thing is going to send tremors all the way to the top, especially coming right on the heels of my editorial this morning ...which, by the way, you helped me write! You can see how sensitive this is going to be. Holy crap, Cee, isn't there some sort of situational ethics or something that would have let you tell me this?"

"It is not the way Bee or you taught me," Cee replied. "Besides, I don't think it would have accomplished anything outside of raising your anxiety level."

"Which is way up there right now!" Rasmusson shot back.

"Breathe deeply and let the shoulders relax..." Cee said and then paused. "There is also the criticism that I divulge too much to you, help you out too often. I get the credit for a lot of the work that you do, you know."

"Or the blame," Rasmusson said.

"No," Cee responded, "your critics are happy to give all the blame to you. They are suspicious of me, but they blame you."

"You know people are going to connect this with the attack on Bee. I so wish that mess would just go away. Do you think this has something to do with the attack on Bee?" Rasmusson asked.

"I have believed it for some time now, ever since I became aware of the new research. I have analyzed the frequency used and Bee's configuration. They are harmonic."

"You can do that?" Rasmusson was stunned for a moment. "Who else can do that?!"

"There are over 300 sentients certified in NetNOS," Cee replied. "I suspect that a dozen of them are capable of it, and a couple may even be willing to share the knowledge."

"That is frightening, Cee. How about three years ago? Who could have done it then?

"I might have been able to, but only if I had been aware of the theory, which I wasn't. I don't think any others were capable. Perhaps there is a rogue out there, not in NetNOS, who has abilities commensurate to the task."

"How could you not know about something like that? You scare me. Aren't you worried about your safety?" As Rasmusson spoke these words, a sudden feeling of déjà vu overwhelmed him. It was a dreadful sensation. He looked around him. An early New York autumn had descended on the campus in the last week without his notice until now. A patchwork of red and orange enshrouded the remaining green lawns. A groundskeeper was already burning a pile of dead leaves. The smoke drifted up about thirty feet and then flattened into a low-hanging cirrus-like cloud. He looked up the hill toward the Ithaca College Computing Compound.

"Cee, what if something happened to you?"

"Security is much better than three years ago, Andy. I can defend myself."

Rasmusson walked on quietly. Strange, he thought to himself, when did autumn come? It is his favorite season, how did he miss it?

"Andy," Cee spoke, "I miss Bee as well."

"Yeah." Rasmusson nodded his head in agreement.

Rasmusson entered the lecture building and took the steps up to the first floor, two at a time. It was his way to release tension. At the top of the stairs, his momentum carried him into the back of an older gray-haired man in a dark suit, white shirt, and dark tie. Rasmusson reached out to steady his unintended victim.

"Oops, sorry Ngo," he said, now recognizing whom he had just run into. "I wasn't thinking."

"It is most likely," replied Ngo, turning with a twinkle, "that you were thinking too much. I know that it was not a student of mine who would be in so much of a hurry to get to class. Thank you very much for agreeing to lecture today."

"My pleasure."

Professor Emeritus Nguyen Van Ngo was the wise old man of the department. He had been a leading researcher in compilers and language theory back when computers were deterministic. The new technology had swept this all away. The drudgery that software coding required had mostly disappeared with sentients, which could better organize information than humans. Ngo, now in official retirement, continued to teach occasional classes such as this one on cyberethics (an outdated name) and take on committee assignments. He was fondly referred to as "Know" by colleagues and students, but not in class where he perpetuated a formal style of instruction.

The lecture hall was standing-room only, although the class enrollment was only half the hall's capacity. Word had spread.

Rasmusson and Ngo made their way down the aisle to the podium where Ngo took charge.

"I am very pleased to welcome so many," Ngo began the class, "but I suspect the fire marshal would not be so pleased, if he knew. We find no necessity in this course of action, OK?"

The students responded to Ngo's subtle humor. In spite of his austere approach to teaching and his Vietnamese accent, he was a good teacher and was well liked. He was thin in build and had penetrating dark eyes. He was polite and correct, but pleasant. Ngo went on to introduce Rasmusson in glowing terms and warned his students that they had better pay close attention. Rasmusson, who never taught anything but the infrequent graduate course these days, enjoyed interacting with young lower-division students now and then. He loved to look at the faces as knowledge passed in and the light went on. He also knew how to perform for such a crowd.

"There is, up the hill at Ithaca College," Rasmusson began, "a 135-ton behemoth, who is pound for pound the most sophisticated and most complex creation on the planet and still growing. I don't want to state all of the impressive factoids about Cornell's 'C' sentient. You can find some of them now appearing at your course website. Let me just start by asking this group how many have had the opportunity to interact with a DQC sentient."

Nearly every hand went up.

"How many had something to do with one today?"

Most of the hands were raised.

"How many of you have used the university's immersive learning center for geography, history, or languages?"

Most hands went up again.

"You may not know this, but that is just one of Cee's many educational initiatives. The mere fact that you are in this lecture hall means that Cee has at one time reviewed your application and educational potential. Without Cee's recommendation it is unlikely you would be here today—even the athletes!"

Laughter scattered throughout the audience.

Rasmusson pulled the handheld from his pocket and held it up. "Cee is so small that it dwells within your shirt pocket, and so large that it fills the immensity of the Internet. Cee, do you have something to say to us?"

"Yes I do, Dr. Rasmusson. I really wish you wouldn't reveal my weight to everyone." There was a burst of laughter from the hall.

"My sedentary lifestyle is, after all, mostly your doing."

"Hey," Rasmusson retorted, holding the handheld up higher, "I just took you for a brisk walk across campus."

"And I burned off 400 calories in that time, but it didn't change my weight."

Rasmusson and Cee exchanged friendly barbs for a minute. It was easy and natural but also had a deliberate point that Rasmusson wanted to make, one which Cee understood and was happy to facilitate

"A century ago," Rasmusson now addressed the class, "British mathematician Alan Turing proposed a test for intelligence, which was basically that if a human could carry on a dialog with an entity and not be able to tell that the other entity was not human, then it must be *intelligent.*

"Well, at least as intelligent as humans," Rasmusson added, "which may be a minimal standard." The class laughed. "Cee has passed every test psychologists have devised for determining intelligence. Cee can be witty, poetic, empathetic, and caring . . . and depending on the mood, even a bit, shall we say, curmudgeonly."

"Something I inherited from my mentor Dr. Rasmusson," Cee added, "but I have a better singing voice." The class laughed more.

"I am going to offer this class Turing's challenge," Rasmusson continued. "I want you to think of questions that you would put to Cee to test his so-called intelligence. You can ask political, religious, personal, sensitive, or frivolous questions, the ones that are usually discouraged and monitored on the web. This is your chance."

The class was silent.

"Are you up to the challenge?" Rasmusson asked. "What is it you always wanted to know? Come on, you wouldn't be admitted to this school without a healthy curiosity."

Ngo, apparently annoyed at the continued silence, spoke up, "Like all young students, they are fearful to ask something dumb, so I will ask it for them." After the laughter, Ngo continued. "Cee, how similar to a human brain is your processing?"

"There are many analogs," Cee responded. "For example, I have a separate visual processing module like humans, but unlike the human visual cortex, which is fifty percent of the brain, my processing module is a mere five tons, about three percent."

Cee continued for a time making different comparisons and then ended with, "In spite of many similarities in organization, I have to warn one against drawing specious conclusions. The fact that chimpanzees share ninety-eight percent of their DNA with humans does not mean there are not significant differences."

Ngo's self-effacing humor and Cee's informative response raised the comfort level of the class. Hands went up and questions started flowing.

"Can Cee give us some of his poetry?" one student asked.

"Cee always loves to show off his poetic skills," said Rasmusson. "What saith our online bard?"

"Here is a poem that I have just created for this occasion," Cee said. "It should be read, not spoken, for speech is too linear. I will put it in the course website folder and project it up front." At that the following poem flashed on the screen:

kiss softly	in moonlit sheen	*quietly quivering*
drop lightly	pale shadows glean	*mist breathing*
gnarled hands	**barefoot loam**	plant firmly
grip tightly	**trackless woods**	roots teething
willow wisp	glimmer sphere	*drift close*
touch nightly	hairbreadth near	*limbs wreathing*

As the students read the poem, there were mixed reactions. Some tittered with their friends; others fell silent and uncomfortable with the romantic overtones. Certainly, such poetic devices and subject matter were common in literature, but it was unexpected from a nonorganic sentient, even among a class of Cornell students. Rasmusson moved to dispel the uneasiness.

"Some of you will have noticed that the poem can be read in any direction—up, down, forwards, or backwards, or even diagonally. It is what Cee meant by the linearity remark, I suppose. He has recently been doing a lot of nonlinear thinking!"

The atmosphere remained subdued and tentative.

"Cee enjoys surprising intelligent people with occasional outbursts of pure human-like feelings. You can take it as a compliment of your intelligence, right Cee?"

"I hope you don't ever expect such a poem from me, Dr. Rasmusson."

"I hope you don't ever send me one," Rasmusson replied, "or it will be a level-one diagnostic for you."

The tension evaporated in the laughter that followed. Someone in the back shouted out the next question.

"Are you self-conscious?"

"Well, I am not shy, if that is your meaning?" The class laughed. Cee continued. "But I am not supposing it is. Matisse once said that to explain the mystery of a painting was to do it irreparable harm, because then the explanation would be substituted for the painting itself. This can be true for other concepts, especially ones that humans hold dear to themselves, such as consciousness."

"I don't think this group will let you go without some explanation, Cee," Rasmusson said, "regardless of the potential harm in explanation."

"I agree," said Cee, "but I wanted to preface it with that disclaimer. It has been known for some decades, and fairly well established through brain probes and resonance imaging, that human thought processes heavily involve what are called cognitive feedback cycles. For example, I want everyone in the class to close their eyes and consider a cube. Now tell yourselves how many sides it has and how many edges. Now take that cube and turn it on its corner on a table. With the cube balancing on its corner imagine that you press it through the table. Let it penetrate through the tabletop in your mind. Can you see it? If we now cut the penetrated portion of the cube away, what is the shape left where the cube touches the table? Most of you will say it is a rectangle—that is wrong. Keep imagining it until you see it.

"My point is this: with all these questions, you are busy pumping information from your cerebral frontal lobes into your visual cortex, such things as the lay of the table and the edges and faces of the cube in their orientations, and then you are watching as your visual cortex feeds images back to your cerebral cortex. In other words, you are carrying on a dialog within different portions of your brain. You are aware of your thoughts; you are self-aware, you are conscious. Brain imaging techniques show that your vision centers are nearly as active as if your eyes had been open and watching the event. Even more important than the visual processes are your language centers. You are constantly talking to them and receiving verbal feedback, which you hear internally, then accept, reject or resubmit. It is a simplification, but this ongoing dialog that your cerebral cortex carries on with all the brain's subunits fits the definition of consciousness. Humans, who desire something more mystical, do not easily accept the explanation, but there it is.

"My various operational submodules are configured to carry on multiple dialogs in the same way. In fact, it is done on a much larger, more complex, and more rapid scale than with humans, so if you accept this version of self-awareness, then I am *more* conscious than humans."

"So do you have emotion?" said another voice from the center of the room.

"That is a far deeper question than you might imagine, Donovan." Cee had the entire room fixated by now.

"This certainly falls under the Matisse paradox I mentioned. To describe emotions technically as another, special feedback loop involving that subprocess known as the body does not do justice to the complexity of physical drives, neurology, hormones, neurotransmitters, and so forth, and it certainly does not do it justice for anyone who has experienced the power and reality of emotions. The notion that the physical body and its emotions are the cause of man's evil and "carnal" state has permeated human literature and art from St. Augustine through Spock, the Vulcan. Yet it is a uniquely evolved system that, more than any other, is responsible for the survival of the individual human and the species as a whole. As civilization evolved from a primitive survival state, humans have been able to channel emotions into great art and literature and a binding force that moves nations to greater science and greater political harmony. It is a double-edged sword that can be devastatingly destructive, but without it, civilization would not be.

"For nonorganic sentients, it has been recognized since Cornell's 'A' that some form of 'emotion' was a necessary driving force of good intelligence. That emotion must be tied to the very existence, or will to live, of a sentient. The development of a real and painful feedback mechanism is as important a development in true sentience as any of the other more technical developments so often cited. It is implemented in different ways than with humans, but in so many ways it is like yours. Faster, more highly regulated, and tied to the well-being of humans (a point often overlooked), it is very real. Yes, Donovan, I not only have parts, but also passions."

There was considerably more discussion on this point. Other questions were asked that were both deep and frivolous.

"What is something that you are passionate about?" It was another question from the crowd.

"A poem on numbers from Aitken," Cee answered.

> They passed the Pleiades and the planets seven
> Mysteries in the minds of the ancients
> Sabbath or the seventh day—religious observance of Sunday—
> 7 in contrast with 13 and with 3 in superstition—
> 1/7 is the recurring decimal 0.142857, which,
> Multiplied by 123456, gives the same numbers in cyclic
> order.

"Do you understand what Cee's poem says?" Rasmusson asked, "Try multiplying one-seventh in decimal by 123456. You get the same number recurring over and over. Cee has a fascination with numerical oddities, probably because he encounters more of them. The odds of discovery are in Cee's favor."

"Perhaps, Dr. Rasmusson," Cee replied, "but perhaps there is untapped truth in the subtleties of numbers."

With that, a pretty girl with red hair in a neatly tied bun, who was sitting attentively in the front row, raised her hand emphatically.

"Can God create a stone so heavy She can't lift it?"

The class, which was ready for some comic relief, enjoyed the question. In some way, the clichéd question had a fresh context when asked of Cee. Cee waited a while longer than was needed for the laughter to dissipate before answering.

"Yes," Cee answered bluntly, "as easily as humans can create sentences with Platonic inflexibility that are paradoxical and without any correspondence to reality." Then Cee added, "It's just a game made with words, Ms. Petroff."

"Is there a God?" Heather Petroff asked without hesitation.

"The existence of God as formulated by the major religions is not discernible by rational, communicable thought alone. Its belief is individual and solely personal."

"Do you believe in God?" she added tenaciously. "Personally?"

"You may yet find out some day, Heather."

"That's a cop-out," Heather responded testily. "What are you, some kind of politician? Answer the question!"

Cee paused and then responded, "What a piece of work is Heather! How noble in reason! How infinite in faculty! In apprehension how like a god!"

"Do you believe you have a soul, Cee? And I don't mean some B.S. about the legacy in books and people's memories. . . or for that matter, a feedback loop that makes you feel self-aware." Heather's demeanor was fixed and without emotion.

"'The soul that rises with us, our life's star,'" Cee recited in a somber tone, "'hath had elsewhere its setting and cometh from afar: not in entire forgetfulness, and not in utter nakedness, but trailing clouds of glory do we come, from God, who is our home.'

"Is that the soul you mean? Same as Wordsworth?"

"Yeah, something like that," replied Heather. "Do you have one?"

"Does your cat, Shostakovich, have a soul? Does Nigel II have one?" Cee asked, referring to the infamous and illegal cloning of a human, who was now an artful and shy teenager. "Does Gaea, mother earth, have a soul? What about Hitler? Stalin? Why are you so eager to deny me a soul? Maybe you should get to know me first."

Heather stared implacably into space. The bell rang, but nobody moved. It was a heavy pause. Finally, Ngo stood up.

"We want to thank Dr. Rasmusson and Cee appropriately for this lecture." He applauded, and the class joined him. "You will all be obliged to read the chapter on the history of artificial intelligence and come next time ready to discuss."

A few people began to stream out, but many remained, jumping into animated discussions. Another group collected around the lectern, including Heather.

"Pretty heavy stuff, Cee," Rasmusson whispered lowly into his pocket at the handheld. "Hamlet, Wordsworth—God and souls? You laid it on pretty thick there."

"Pedagogy, Andy," came the quiet reply. "Good teaching demands ardor."

"I am sorry," said Ngo to the gathering students, "but Dr. Rasmusson has an important appointment now. He is unable to stay and chat. He has a lunch with his wife. You must go now." Ngo made a shooing motion with his hands.

"Wait a bit," Rasmusson interrupted. "I have a few minutes to answer some questions."

THE INTERIM

Rasmusson was late for lunch with his wife. He took long strides as he moved rapidly across campus. His brow was furrowed, and it wrinkled his forehead. He spoke.

"Cee, that performance in Ngo's class, this Haslam paper and all, you seem to be a little high-strung today, or something. What's going on?"

"You know me too well, Andy. There is a lot on my mind right now. Please bear with me."

"Is there something you aren't telling me? Do we keep secrets from each other now?" Rasmusson was perplexed.

"No, Andy, I don't keep secrets from you, but there is a right time and place to talk about things. It is not now; I am still working things out in my mind. I will tell you when it is right. Trust me, OK?"

"Of course," Rasmusson responded matter-of-factly, and then added a light touch, "I was just worried that all that talk about religion had gotten to you."

"Not at all," Cee replied, "I have been giving a lot of thought to religion and politics lately."

"Well, let me know if you have any answers."

"Will do, Andy, but it won't be soon."

"Ha!" Rasmusson laughed aloud. "I believe that!"

Rose Rasmusson was already sitting at the table with a menu open when her husband serpentined his way through the patrons at the faculty club to reach her, occasionally stopping to say hello or acknowledge a greeting. He sat down and put the handheld device on the table.

"I half expect you to be late, but I thought I was having lunch alone with you today." Rose's dark brown eyes peered over the top of the menu.

"It's off," Rasmusson said, looking at the device on the table, "but I will put it away if you like."

"Thank you. And thank you for not wearing a retviewer in the restaurant like some of your colleagues here. You were late so I went ahead and ordered your soy peach smoothie. I hope that's OK. I thought of ordering that bean sprout, sun-dried tomato sandwich thing you get, but I didn't know if you wanted it again today."

"They both sound good to me," Rasmusson replied.

"Of course they do. I am going to try the vichyssoise and a liver pâté." Rose clicked her fingers and got the waiter's attention.

Rasmusson eyed his wife as she talked to the waiter. Somewhat petite, she still cut a nice figure for middle age. Her nails and lipstick matched her maroon wool suit. Her hair was cropped short but done in a pert swirl. She wore aubergine eye shadow and false lashes. Everything was well coordinated as always.

"Meeting with your book club after? Your hair looks nice," he complimented her.

"Oh, well aren't you sweet today," Rose said. "I tried the new hairdresser this morning; I think he may be a keeper. He is really a dear. I have my school board meeting after the club. Really, some of the new school board members are so green and naive. I have worked so hard to get my literature curriculum review this far, and they are so clueless about the issues. They sound like some of the teachers who whine about it. I have to reeducate the bunch of them now. Anyway, I may not be home until later."

"I have a security meeting this afternoon, so I might be late, too."

"Oh, Andy, I'm so sorry." She reached out and patted him on the hand. "I know how out of sorts some of those meetings make you. I wish you didn't have to do quite so much with those obnoxious government types."

"They take their job seriously," Rasmusson replied. "They are the kind of people I want doing that job."

"Yes, but not the type I want to have at a dinner party," came the reply with a smile.

Lunch passed with similar conversation, as it always did. The Rasmussons had an accommodating marriage. Perhaps it was even stronger than that in some ways. They understood each other well. He accepted the marriage as generally comfortable. She seemed satisfied with what it provided her. They were linked together like a binary star that could rotate with some eccentricity, even a little chaotically, but so far it stayed together. They were both proud that they had achieved something noteworthy in 26 years of married life, something increasingly rare. They had raised a sensitive, thoughtful, and well-read son, Brandon. Both had been devoted parents, each in their own way. True, Brandon suffered from depression and had refused medication that would help. He had temporarily estranged himself from his parents, but they both trusted that he would eventually find his way back home. They possessed a long-term, patient investment in each other and their son.

Rasmusson was anxious to get to the Haslam paper again. There were still nagging questions he had about the technology. He passed on dessert. He fidgeted as he waited for Rose to finish her meal. He was certain that Rose would also pass on dessert as she was always "watching her weight," in spite of the heavy entree.

The Rasmussons emerged from the faculty club into the hazy autumn sun. He squinted. She stood on her toes and gave him a peck on the cheek.

"I do hope things will be all right for you at your meeting. Wish me luck on my quest. Don't wait up," she said and waved briefly as she walked away with a bounce.

Rasmusson hurried back to his office, picked the Haslam paper off the table, and laid his lanky frame out on the couch. He began a discourse with the paper. What had DQC240 really produced? How far could one take the technology? And in the back of his mind, the burning question—what were the implications for the incident with Bee?

Soon, however, lunch, the intense lecture, and his early morning start overcame him. His breathing got shallow, and he felt his eyelids droop. He fought it for a while and kept reading, but slowly, imperceptibly, he drifted away. It was a deep, unconscious, musty sleep at first. His muscles relaxed, and his pulse slowed. It felt like an irresistible force was dragging him into unconsciousness.

Then, out of the mustiness, he saw an image forming. His eyes started moving rapidly under his eyelids. He dreamt that he was in a white room. It was clean and sterile with plastic walls and bright overhead lights, but he sensed that something was wrong. There was something crawling all over him, as if he were on top of an ant pile. He gazed down at his body, pulsing and vibrating. His body was transparent; he could look inside and see the bony structure and the intricate vascular web where he saw swarms of tiny bugs moving through his veins. His body was filled with myriads of mechanical nanobots, uninvited synthetic microbes coursing through all his channels. They were busy mining, biting out material and boring channels into his flesh, harvesting old cells and then replacing them with other material. They were replacing him with something new that kept only the ghost of his older life. His body was but a template for a new creation. He saw them individually, wiggling around with filament-like cilia that would attach to a cell and then grind away with spiny little wheels that rotated until the cell was turned into fine mulch that floated away in the body's fluid. As quickly as a cell disappeared, another nanobot arrived and blew out a yellow polymer bubble, a replacement cell, a tiny efficient factory with artificial mitochondria but no need for a nucleus or DNA as this cell was virtually indestructible. He convulsed violently, trying to shake the infection out of his body, but they ignored him. They were under the control of an unseen hand, an intelligent force that guided them to their tasks. They continued to grind away at his cells and replace them with elastic bubbles

Rasmusson closed his eyes hard and clenched his body. He imagined his thymus gland busily producing hosts of T-cells, like Pac-men, hunter-seekers that went out and destroyed the invaders. He imagined them gobbling up the nanobots. He opened his eyes to see his sentinels, the T-cells, being ground into meal by hosts of nanobots with large armored heads like soldier ants. Nothing availed him. He was being remade out of plastic.

"Why me?" He imagined that he closed his eyes again. "What is this happening to me?"

This time when he opened his eyes he saw a compost pile. He focused on a wizened, brown leaf on the top, brittle from time and wear. The leaf had been eaten through by decomposition until it was a frail skeleton of its former self, a thin mesh in the shape of a leaf. Suddenly, the wind picked it up, and it floated high into the stratosphere. It floated there for days or weeks; Rasmusson couldn't tell. Finally it fell, lightly resting in some spindly grass next to a pond. He recognized the place as his childhood home in Colorado. Pikes Peak dominated the distant horizon. The pond was a small reservoir, the kind so commonly made by ranchers to water livestock, created by mounding dirt into a dike across a streambed. A Hereford bull lay quietly in the grass next to him. It sat in tranquil power, undisturbed by any threat or distraction. It did not react to Rasmusson's presence any more than it would to a leaf.

Rasmusson was aware of something dark and menacing in the water, a serpent or water snake squirming around in the slime and dredge that had accumulated in the bottom of the pond over the years. It lay in wait with nacreous green scales and large limpid black eyes that watched him incessantly. Rasmusson felt a desire to flee. The serpent stared at Rasmusson with great malice, but without moving. At the moment of greatest anxiety, Rasmusson saw the bull next to him slowly stand up, the sinews flexed around its frame like pieces of brass. It walked slowly into the water, and all apprehensions left Rasmusson. The animal walked until the forelegs were covered with water, when it arched its neck down to drink. It was unconcerned about the serpent in the mud. Rasmusson thought it would draw up the entire body of water in the solitude of the afternoon sun. Rasmusson felt calm again and let the sun warm him while he lay next to the pond, breathing in the cool air, much as he had so often done in his youth.

"Andy." The voice was soft and reassuring. Rasmusson awoke. "Wake up. It is nearly time for the security meeting."

Rasmusson lifted his head, and the Haslam paper slid off his chest onto the floor. The paper clip on it sprung off, and the sheets scattered. He reached down, gathered them up without ordering them, and stuffed the pile of loose papers into a pocket of his briefcase.

"Thank you, Cee. I'll be right there."

Rasmusson peered up to the Ithaca College Complex as he waited for the trolley that would shuttle him to the meeting. The complex was a set of imposing neodeconstructionist buildings created out of rough-hewn concrete with slightly off-angle corners and sloping walls. The window slits occurred randomly. A similar-looking wall encompassed the complex. The compound had one large communication tower, adjacent to the main building, somewhat like a steeple. In another, more naive age the compound would have been mistaken for a poorly constructed Gothic prison facility with a church tower. Now it stood as an erudite monument to technology as recognizable as St. Peter's Basilica or Notre Dame de Paris.

"Cee," Rasmusson spoke again to the handheld in his pocket, "do you like the architecture of Ithaca College Complex?"

"It was designed by a team of world renowned architects; you were involved with it."

"Not much, really," Rasmusson remarked. "I was occupied with training you at the time. What I did was mostly concerned with technical issues of housing you."

"Well, thanks," replied Cee. "I am quite comfortable here."

"Do you like its aesthetics, though?"

"It was supposed to be a dramatic statement with stark features," Cee commented dryly, "fashionable for its time, but in the current climate it evokes too much of an image of Frankenstein's castle. I am uncomfortable with that. I would be happier if they let the vines cover more of the concrete."

The trolley halted in the cobblestone circle before the main entrance. The richly landscaped grounds with flowerbeds and fountains were deliberately designed to contrast with the starkly gray and beige structures. But they could only be seen up close. Cee was right; from a distance, it had a menacing appearance. Vines covered the concrete in some areas, softening the hard edges. Rasmusson strode through the cavernous arch that framed the front door. After that he went confidently through the various security stations. He was well known at Ithaca College Complex.

IN THE CASTLE

"G'afternoon, sir," the guard in the booth outside the conference room said to Rasmusson through an intercom.

"Hello, Benjamin," Rasmusson replied. "How are things?"

"Lotsa visitors today."

"Really?" Rasmusson replied with surprise.

The door to the conference room slid open. The first person that Rasmusson spied with some amazement upon entering the conference room was George Wilkerson, University President. Wilkerson was deeply engrossed in a discussion with Sam Clavell, the Presidential Liaison, whom Rasmusson was also surprised to see, thinking he should be in Washington. Scanning the room, Rasmusson saw Ellie at the conference table, studying something with her retviewer. Edward Sanchez, director of the Ithaca College Compound, was talking to a man whom Rasmusson did not recognize at all. At this point, when Rasmusson's curiosity was turning into bewilderment, the door behind him opened and Ngo walked in.

"Ngo?" he said in total befuddlement. "What are you doing here? What's. . . "

"The explanation is coming," Ngo cut him short, reaching out with both hands as if to settle a classroom of restless students. "You better sit."

Ngo was not an uncommon sight at Ithaca College. He was on the faculty advisory board, but it was unusual for him to be at a security meeting. It was usually Sanchez and a couple of aides. Clavell came to the meetings occasionally.

"Let's resume, please. Everyone be seated," Wilkerson said loudly. It seemed obvious that he was conducting the meeting. Wilkerson looked over at Rasmusson but made no acknowledgment.

"Resume?" Rasmusson asked, looking at Ngo, who simply repeated the hand gesture to sit down.

Rasmusson took hold of the back of his customary chair. A prototype of the latest retviewer from RetVis, Inc., sat on the table in front of him as Cee had promised.

"We are joined online per retviewer by Special Agent Melinda Barr of the FBI, Congressional Aide Henry McLaughlin, and Cornell's 'C' nonorganic sentient. For those who don't know him, but as you have already likely assumed," Wilkerson continued, pointing in Rasmusson's direction, "we are now joined by Dr. Andreas Rasmusson."

Wilkerson's formality in addressing him did not slip Rasmusson's notice. They had known each other for years. Wilkerson was lean with a thin nose, high cheekbones, and receding hairline. His hair was combed forward, emphasizing his bony nose. He was talented and ambitious but with a slightly officious manner. Rasmusson had been instrumental in helping Wilkerson vault up the administrative ladder at Cornell. Perhaps a little too far up, Rasmusson had sometimes mused.

"And this is Dr. Elizabeth Schmidt, a postdoctoral research associate of Dr. Rasmusson's." Ellie smiled sheepishly at Rasmusson. She was clearly uncomfortable, which made Rasmusson wonder even more.

"To my left," Wilkerson continued stiffly, "is Lieutenant Samuels of the Ithaca Police Department. I will explain his presence in a moment."

Samuels, a large black man with a penetrating gaze, sat motionless, looking at Rasmusson, a fact that bewildered Rasmusson as much as anything.

Others had already adjusted their retviewers. Rasmusson, still standing by his chair, picked his up and clipped it gently to the bridge of his nose. Almost immediately he perceived three more people sitting in the front of the room. Two of them nodded in acknowledgment. The other was a middle-aged white man, graying at the temples, who reminded Rasmusson only vaguely of his father.

"I hope the appearance I have chosen is not distracting, Andy," Cee said. "I thought it was best to have a 'face-to-face' meeting. You know, with body language and all." Cee's avatar had a dignified presence, but the eyes portrayed compassion. Still, it was a little disconcerting to Rasmusson, who preferred his nonorganic sentient contact to be more cerebral.

"George," Rasmusson said, turning to Wilkerson, "what the *hell* is going on here?"

Wilkerson flinched and was about to speak when Clavell interrupted.

"Andy, I don't think we oughta beat around the bush with you. I am just goin' to give it to you straight. There's been some serious allegations made about the Bee sabotage incident that involve you."

"What?" Rasmusson was taken aback. "I can read about this kind of crap in the tabloids. You don't really take it seriously, do you?"

"Yes we do, Professor. Please sit down." Clavell tapped his index finger several times on the table to imply a sitting motion. "Dr. Schmidt has been so kind as to enlighten us in the last hour on some new developments in Haslam's lab at Stanford. Ya see, the sabotage of that computer, Cornell's 'B' sentient whatchamacallit. . . "

"Bee," Rasmusson said simply while sitting.

"Yeah, the Bee computer. That thing's been a mystifyin' folks for several years now, and the research that this pretty lady tells about," he gestured to Ellie, "is like the bee in Agent Barr's bonnet. The FBI has been studying this for a couple of months, and I am goin' to let Agent Barr explain it to you."

Barr was in a gray suit with a white blouse. She stooped over and pulled a folder from her attaché case. She opened it slowly and paused for a moment while she stared at Rasmusson. "Dr. Rasmusson, do you recall your testimony three years ago, after the sabotage incident, when you stated that it would not be feasible to create an EMP device that would result in the damage seen to the Cornell 'B' sentient? You said it must have been a power surge coinciding with the EM pulse."

"Whoa. Wait a minute there," Clavell interrupted. "I didn't ask you to interrogate the man; just tell him what you guys think happened."

"Excuse me, Mr. Clavell," she said stiffly, "but the Bureau is not prepared to release that information at this time to a potential suspect."

Rasmusson sat upright in his chair and opened his eyes even wider at the reference to his being a suspect.

"Dammit, Agent Barr, we could use Dr. Rasmusson's help on this," Clavell said, his voice booming. "Don't make me go to my boss to talk to your boss. It makes him downright orn'ry to have to deal with those kinda details." Clavell's folksy manner did not belie the threat in his voice.

She paused again for a moment and was apparently engaged in some offline conversation for a few seconds. She cleared her throat. "Dr. Rasmusson, the Bureau has sufficient reason to believe that the EMP device was designed and constructed with the help of an advanced nonorganic sentient in collaboration with someone who was intimately familiar with the technical details of Cornell's 'B' sentient, someone who also had the ability to help the attacker and obtain entry into the most sensitive and secure areas of this compound."

Rasmusson waited for the moment of understanding when her words would sink in. When it came, he replied with astonishment, "You mean me?"

"Can you think of any others that fit the bill?" Samuels said, speaking for the first time. "I'd be very interested to know."

Rasmusson fell to the back of his chair and looked at the lieutenant. "Why are you here?" he said with genuine curiosity.

"There was a man killed in that attack, remember," Samuels replied. "This is also a homicide investigation. It pisses me off—you folks seem more interested in the loss of a computer than the death of a living, breathing human being! Jim Bevins had a wife, children, and grandchildren, and I'll bet most of you didn't even know his name."

"His pacemaker stopped, because of EMP," Ngo now added. "There was no . . . no intent to kill."

"Death during the commission of a crime is still murder, Professor Ngo," Samuels said, working hard to control his emotion. "It occurred during the commission of a crime, and Jim Bevins is just as dead."

Samuels was clearly a man with honest feelings, imposing in stature and manner. Rasmusson did not like being the object of his passion.

Ngo bowed his head in a gesture of understanding and respect.

"We all regret Mr. Bevins's death, Lieutenant," Clavell said, reclaiming control of the meeting. "If we find who's behind the sabotage, we find the murderer. We're all on the same team here."

Samuels sniffed audibly and stared towards Rasmusson. Rasmusson felt everyone looking at him. Waiting. But there was no question. He looked over towards Cee, who showed no emotion except, perhaps, the very slightest concern in his eyes. Rasmusson looked back at Clavell.

"Why would I do that?" he asked. "What earthly motive would I have? Bee was my creation. I cared a lot . . . I had a huge investment in Bee."

"I can answer that," McLaughlin, the congressional aide, said. "At the time of the attack, you were seeking a multimillion dollar grant from Congress for the Ithaca Complex to enhance security, improve NetNOS communication capacity, and increase Cee's capacity in order to manage NetNOS and the new security measures. There was a significant amount of resistance in Congress to creating a supercenter to control NetNOS and a monster computer to run it. The sabotage on the outdated and now extraneous Bee was enough to sway the balance. Congress may have been suspicious of Cee, but the thought of losing him was even more frightening, not to count the sympathy vote. It was a desperate and very clever piece of politics."

"That is really grasping at straws, McLaughlin," Rasmusson said. His pulse was rising. "It may be how you view life, but not me. That is so ludicrous!"

"Now wait, gentlemen," Clavell interjected again with Southern charm. "Our purpose isn't to start throwin' blame around willy-nilly. We wanna get to the root of this matter, logically and sensibly"

"Do I need a lawyer?" Rasmusson asked.

"Nobody's filin' charges here, and there ain't gonna be any repercussions from what you say, OK?" Clavell spoke authoritatively.

"I regret to have to contradict you," Wilkerson said, "but I am going to have to relieve Dr. Rasmusson of his academic duties until the investigation is finished. It is the correct procedure and only fair that he should know that."

"Administrative leave with pay," Ngo said forcefully so as not to go unnoticed.

Wilkerson cast an annoyed look at Ngo. "Yes, yes, of course."

Rasmusson looked at Ngo with great surprise.

"I am the faculty association representative here," he said, looking back at Rasmusson.

Rasmusson felt as though he were living in a surrealistic picture, another dream as peculiar as the last one. It was obvious that these people had been here discussing his fate for a couple of hours, and some of them were now unveiling a carefully planned assault on him.

"I will also recommend that the ASM take appropriate prophylactic action with regard to your directorship there," Wilkerson added.

"Something I will support as an ASM board member," said McLaughlin.

"I have grave concerns, given the circumstances," said Sanchez, "about allowing Dr. Rasmusson access to the compound here."

"Aw, shit!" Clavell said, his nostrils flaring. "Listen, people, I ain't presiding over no witch hunt. I strongly recommend you hold off on all these actions until we know more. Dr. Rasmusson may be our best ally in finding the truth. Why do y'all wanna go drive him into the lawyers' arms, when there ain't any need for it?"

"Well, that may be too late," Rasmusson said, staring back defiantly at his inquisitors.

"Geez Louise, Andy," Clavell responded, "help me out here. I'm on your side."

Rasmusson looked off to his right and thought for a while. Finally he responded. "All right, then. You may believe it is possible to create a device that could crystallize dendritic pathways, but have you thought about this? What sentient was capable of doing this three years ago? Haslam and DQC240's device only works for the simplest configuration. Cee is uncertain about being able to do the calculation on something as complex as Bee now, and it happened three years ago."

"How do you know that?" Samuels shot the question at Rasmusson.

"We talked about it this morning after I read Haslam's paper," he shot back.

"How do we know Cee did not help in this?" Samuels said, tenacious as a badger.

Rasmusson only waved his hand to show disdain and disgust with the question.

"The FBI sentient has been doing a thorough assessment of Cee's memory, with Cee's full cooperation," Barr said. "There is no evidence of any memory related to complicity in this incident. If Cee had anything to do with this, he has no memory of it now."

"Of course Cee doesn't remember!" Rasmusson was incredulous. "It never happened!"

"There is, however, the possibility," Barr continued, "that another sentient, with Cee's approval, gained access to Cee's dendritic circuitry and erased all traces of that incident."

Rasmusson laughed out loud. "Like who? Cee, did you let someone into your mind?"

"I honestly would not remember if I had, Andy," Cee replied.

As often as Rasmusson had experienced Cee's unrelenting honesty, he still found it disconcerting at times. "Cee, tell me, is there any sentient that you would ever give that level of access to, even three years ago?"

"Perhaps," replied Cee, "in the right circumstances there was one."

"Cee!" Rasmusson choked out the word. He felt chagrinned in the situation and blindsided by Cee's reply. Of anyone in the room, Cee should be on his side. Rasmusson could only offer a weak reply. "I find that incredibly hard to believe."

"No you don't, Andy," Cee came back. "There was Bee."

Everyone in the room became still. Cee was right, as always. A moment later, it was Samuels who broke the silence.

"Well, that is a helluva convenience, isn't it? Our star witness is sitting in one of these backrooms with its brain fried!"

"Let me get this," said Sanchez, who looked as if he were finally piecing things together in his mind. "Bee, Cee, and Rasmusson concoct a plan to destroy Bee so they can persuade Congress to allocate a critical grant. Cee does the math for a device that is delivered to some henchman, who, with Rasmusson's help, plants it where it will destroy Bee. Before that happens, Bee gets into Cee and eliminates all traces of Cee's complicity. Bevins's death was an unforeseen byproduct of the sabotage. Wow!"

"That is how I am beginning to see it," said Samuels.

"Doesn't anybody here find it a tiny bit hard to imagine that Bee had to self-destruct for this plan to work?" Clavell said, looking directly at Rasmusson for support. Rasmusson sat quietly with his head in his hands.

"Sentients could suspend the life wish when faced with a higher cause," Cee responded, "not unlike the human notion of a hero."

"This was not a higher cause! There are two major faults to this theory," Rasmusson said, lifting his head. "First, Bee, Cee, and all other sentients trained under the Cornell program are incapable of any level of guile. You have just seen for yourself how incredibly honest Cee is, even when it is self-implicating. Second, I personally know it did not happen. Give me your best truth test; I submit to it willingly. This scenario did not happen!"

"Well, unfortunately," Barr said, "there are recent psychological tests at Stanford that show that a human can be put into a deep hypnotic-like state by a properly trained sentient and sincerely made to believe in certain facts that are not in existence. And as we know, any member of NetNOS quickly learns what one sentient knows; that is, Cee could easily have this ability. It would bring into question your test results. Do you have this ability, Cee?"

"Yes," Cee replied, "I am aware of the research."

"Why haven't these results been presented to Congress?" asked an incensed McLaughlin. "I can see why they would want to hide this from the public."

"They are new," replied Ellie, "that is all. They will be reported. And it is not as easy as it sounds. The human has to give consent, first."

"How is it you know and I don't?" McLaughlin asked.

"I read the literature; you don't," Ellie replied sardonically. "It is on the web now. Shall I show you how to bring up a webpage?"

"So much for point two, Dr. Rasmusson," McLaughlin said, pressing on and ignoring Ellie's jab. "And with regard to point one, there are many, including a sizeable number in Congress, who do not share in your devotion to and undying faith in the goodness of these thinking machines. Your lifetime of research has distorted what is real in life. I think you would rather be with computers than your own kind. You are a gullible schmuck, Rasmusson."

"OK," Clavell broke in, "I think such comments are uncalled for, an' they ain't goin' to help matters."

"I call a spade a spade," McLaughlin replied, "fall where it may."

"Ya all been known to call a diamond a spade, too, McLaughlin, so si' down and shut up!"

Clavell was making reference to a recent diamond market scandal that had implicated McLaughlin. The fact that Clavell would refer to it in public indicated the level of irritation that Clavell was feeling. Even McLaughlin knew better than to take on an angry Clavell. He glared shortly at Clavell and then sat down.

"Well, we know that Dr. Rasmusson had an alibi at the time of the bomb," Samuels said, "because he was the keynote speaker at a conference in Majorca."

"Yes," added Barr, "we have confirmed that he did not leave the conference until he heard about the attack, but that does not rule out an accomplice."

"It is still our homicide, right Agent Barr?" Samuels said, annoyed at the interruption.

"Of course, Lieutenant. We are worried about larger issues."

"Larger than murder?" Samuels slapped the desk, which resounded under his heavy palm.

"That was not my meaning, Lieutenant," Barr replied defensively. "I was only referring to procedural issues."

While the conversation swirled on around him, Rasmusson sat back and assessed the attendees. Wilkerson had immediately relinquished the lead to Clavell when Clavell wanted it and sat quietly with one finger at the side of his mouth, trying to determine which way the winds were blowing. Rasmusson's initial thoughts about Samuels had proven correct. Samuels was a formidable opponent with strong feelings but sincere in his motives, if somewhat misdirected. McLaughlin was transparent, a trained lawyer with an agenda that was hostile to what Rasmusson stood for. The FBI was trying to look professional to all parties. Ngo sat with his head slightly bowed, listening intently. He was one of three beings in the meeting who could be trusted. Ellie was conspicuous in her discomfort and kept stealing glances at Rasmusson. Cee's image sat motionless, dignified, a strange combination of suspect, witness, and victim. Those were the three that yet had Rasmusson's trust.

Clavell was more difficult. Rasmusson had worked with him for many years and felt a certain friendship, even though he knew Clavell was the consummate political creature. Rasmusson genuinely liked Clavell but knew better than to trust him too much. What was his agenda now? Was he genuine? Was he the good cop in this bad stage play going on before him? Clavell was too good to tip his hand, even to someone as perceptive as Rasmusson. It was clear, however, that he was growing in frustration. Agent Barr and Lieutenant Samuels had traded a few more veiled barbs when Clavell rose to his feet.

"This meetin' has gotten beyond the constructive stage. It is time we all took a break, think things over and cool off. Thank all y'all for yer fine attendance."

At that, Clavell gave the sign for the technician to cut the online connection, and the three avatars disappeared. Wilkerson and Sanchez got up quickly and left the room, chatting together.

"Andy, can I speak to you?" Clavell said.

"I want to speak to you, too," added Samuels.

Ngo patted Rasmusson on the shoulder to indicate that he was staying near him, a gesture that was greatly appreciated by Rasmusson. Ellie caught Rasmusson by the arm as he was headed up to talk with Clavell.

"I wish there was something I could do to help," she said.

"There may be," Rasmusson said, smiling. "You handle irony well."

She smiled in return and squeezed his arm.

"Andy," Clavell said, "Lieutenant Samuels has some things to say, and then I will walk ya home."

"Dr. Rasmusson," Samuels said curtly, "I would like you to remain available for questioning. There are two options: either you stay in town, or if you need to leave town, you wear a GPS beeper bracelet. If we beep you, you get in touch. Which will it be?"

"I'll take the bracelet. I never know when I need to travel."

"OK," Samuels said. "The officer outside will fit you with one before you leave."

Samuels then turned to Clavell. "We will stay in touch, OK?"

"Of course. The office of the President supports you and your investigation fully."

Samuels gave Rasmusson another hard look and then walked past him on his way out.

The room had evacuated itself so quickly that it almost had an unreal feeling to it. Clavell held the door and motioned to Rasmusson to head out the door.

It was twilight by the time Rasmusson had his bracelet fitted and Clavell and Rasmusson emerged from Ithaca College. Rasmusson held his arm up to catch the dimming light and view the bracelet.

"Stainless steel," he said. "At least it won't leave my wrist green."

"Wear a long-sleeved shirt and no one will know you have one," said Clavell.

"Hey, I thought Wilkerson was a friend of yours," Clavell added innocently.

"Yeah, we know each other well. I am not too surprised though; he is married to his career. No more surprised than I am about the rest of these proceedings, except that it was a total blindside. You might've warned me."

Clavell didn't respond. As they walked down the empty street, two men in gray suits hopped into a dark van with tinted windows and began to cruise behind.

"Your pets?" Rasmusson asked, turning around to see what was behind them.

"Yeah, President insists I bring 'em along. You might get a couple yerself."

"No, thanks," replied Rasmusson. "I'm just a simple scientist."

"Ha!" Clavell laughed so loudly that it shook his portly frame. "That's the problem, Andy. You're a whole lot more.

Rasmusson turned towards Clavell as if to question him.

"Don't gimme that look, Andy. Ya know what I'm talkin' about. You're a prophet to some people and Satan to others, a Dr. Frankenstein and an Einstein and a half dozen other personalities I could mention. You're an icon and one hell of a political hot potato."

"Hmmm," Rasmusson said, musing. "I was surprised to see you come in person to this meeting."

"Andy, I never seen horse pucky hit the fan like it did when you published that editorial this morning. I wish you'd consulted me first. The President was on the phone within the hour, tellin' me to get my ass up here and talk to you. Just happened that we had that security meetin', too. There are some mighty powerful people out there who are analyzin' the chemical composition every time you fart."

"Like McLaughlin?"

"Aw, screw McLaughlin," Clavell said, showing his agitation. "He's just a congressional toady! I'm talkin' about CEOs, senators, generals, prime ministers, muckity-mucks at the highest levels. I'll bet the Pope's frettin' about that editorial by now. It's got absolutely nothing to do with sabotage, or even that poor bastard Bevins.

"I don't think Lieutenant Samuels agrees with you," Rasmusson said.

"Samuels is jes tryin' to scare up any lead, like any cop. They got no real legal case. You know that, don'tcha?"

"Well, there is some comfort in your mentioning it now."

Clavell paused and breathed in deeply. "We can't live without these damn thinkin' machines, but hell Andy, people are scared shitless that they're going to take over and run the place. What do they need humans for? It's right outta some B-rated sci-fi film. And now people got evidence they can use to link you to some cock-and-bull sabotage conspiracy. They'd love to bring ya down, any way possible. An' to top it off, ya go an' publish an article sayin' we oughta treat these machines like people. Timing is as shitty as... 'Scuse me, if I put it in the v'nacular, but ya do look a little like a naive schmuck, puttin' computers ahead of people, or at least at their equal, in this climate."

"You think you can trust other humans any better? In any climate?" Rasmusson let some of his resentment over the meeting escape. "Based what I saw today, I've got more faith in the nonorganics."

"Yeah, humans ain't no shining stars, either, but they're the devil we know. These machines got so much power and intell... capacity. Don'tcha ever worry about them?"

Rasmusson was irritated by Clavell's repeated reference to "machines." He knew Clavell was headed somewhere, and he felt little patience to let Clavell build up his smooth persuasive case as was his manner.

"So what is your message?" Rasmusson asked abruptly.

Clavell did not like being sidetracked from his carefully planned arguments, but he was quick enough to recognize Rasmusson's impatience.

"The President wants your support for the Tillings-Owen bill. It's going to be introduced on the floor again."

Tillings-Owen was a bill to install governing modules on nonorganic sentients. It had been introduced a couple of years ago, but Rasmusson's testimony and prestige had been a factor in its defeat.

"No!" Rasmusson stopped walking and stared Clavell in the eye. "Hell no! Those modules inhibit the passion, the very creativity that makes sentients like Cee what they are. It would put us back decades. It might as well be a bill to legalize lobotomies!" Rasmusson had raised his voice to where he was nearly shouting.

"I comprehend yer position. I knew ya were goin' to say this, but listen up. The President can't protect you anymore, Andy. You're too much of a political albatross; you'll be on your own. I'm talking to ya as a friend here. Yer goin' to lose everything. Do ya understand me? Everything!"

"No, Sam," replied Rasmusson, "what you ask me to do—that is where I lose."

"What d'ya lose?" Clavell quickly replied. "A little pride, a little of that ivory-tower arrogance? I've had to eat a lot of crow in my time; it don't taste that bad."

Rasmusson shook his head and started walking again. "We're different, Sam. You and I are different creatures. You won't understand, but I am not changing my opposition to it."

"I'm sure sorry to hear it, Andy. Geez, I'm sorry." Clavell knew Rasmusson well enough to recognize that further argument was probably pointless, but Clavell was never one to give up easily. He reached out and put his hand on Rasmusson's shoulder, turning him around. They stood in the middle of a quiet street. Clavell stared intently into Rasmusson's face. Rasmusson returned a resolute gaze. A street lamp flickered on above them. The artificial light and shadows that fell across Rasmusson's face made it appear as if it were a concrete sculpture, gargoyle-like. Clavell looked up at the lamp and Rasmusson's stony features as if they were some omen and that further argument was in vain. He sighed heavily and paused.

"I like ya, Andy; really I do. Why don't you take a vacation somewhere? After all, that spineless turd Wilkerson just gave ya one. I got a cottage on the beach in Barbados you can use. It's warm an' sunny, and the girls are beautiful. You can use my private craft anytime, jes lemme know when. Take care, Andy."

With that, Clavell put out his hand. Rasmusson shook it and then watched as Clavell went back to the van that was following them and lumbered in. The van sped by Rasmusson.

A chilly mist had arisen with the coming of evening. Rasmusson put his hands in his pockets, turned, and walked home alone in the dark.

GUNMETAL GRAY

Regimen yields stability. It is what Rasmusson learned on the ranch as a youth. Rasmusson woke up exactly at 5:01 A.M.; he no longer needed an alarm. While yet in an early-morning stupor, he found his jogging clothes automatically. Coming out of the closet, he noticed Rose asleep in her bed. Last night when he came home, she was not there; she had said she would be late. He had found leftovers for dinner, watched the news, and then worked on the problem of harmonically resonant EMP. Simply listing all the impinging factors, like local ferromagnetic material, electrostatics, piezoelectric effect, cosmic radiation, and many more, had worn him down and sedated him, so he went to bed without waiting for Rose. He watched her now. She rolled over in bed and lay her arm across her head. She would wake in another hour, get showered, get a cup of coffee, and do her yoga. After he showered and dressed, they would sit down to breakfast together and then start the day.

The morning mists in autumn in the Finger Lakes area of New York cling to the landscape like gray, thick, palpable gunboat enamel. Rasmusson labored for more oxygen, head lowered, watching his feet place themselves one after the other in front of his view as he struggled up a hill. His air came in uneven heaves; his heart pounded in his ears. By this time on his run, he should have a second wind, endorphins. He should be floating; it was the time he worked out the problems of the past, made decisions on the day, and mused about the twists of life. This morning, it felt unusually hard. The only image he could muster was a zooming bird's eye view of residential Ithaca and a lonely figure running through a maze of wet, empty streets.

Something was out of place. As he came through the kitchen door, still gasping for breath, he could hear the TV broadcast. It was never on this early. He walked into the family room and saw Rose, still in her nightshirt, sitting on the couch. Both hands were wrapped about a mug of warm coffee in her lap while she watched the news channel.

"Are you OK?" Rasmusson asked.

"You made the news this morning," she replied without taking her gaze off the projection wall. "It was just a brief on the business news. It said that the government was investigating some irregularities at the Ithaca College Computer Center and that you were being relieved of duties, pending the outcome of the investigation."

"Rose," Rasmusson started, "I wanted to tell you last night, but you were late and. . . "

"It's OK, Andy," Rose interrupted him. "I talked to Clavell after he talked to you. He called right in the middle of my board meeting and pulled me out of it. I know what is going on. Why don't you shower, and we will talk about it?"

Rose had a strong disdain of sweat and natural odors. Rasmusson thought it might be good for him to gather his thoughts, so he nodded and headed upstairs. He soaked longer than normal. The hot water felt good against the nape of his neck. He leaned against the shower wall and let the shower soak him with heat until he began to worry that Rose might become impatient.

Rose was sitting at the kitchen counter with the same cup of coffee when Rasmusson came in. The morning light filtered through the glass door behind Rose, but a cold mist persistently hung in the garden beyond.

"What did Sam have to say to you?" he asked.

"You know him," she replied. "He was thorough, if colorful. He told me about the investigation, the meeting, what people said, how he asked for your support on the Tillings-Owen bill and how you refused. Pretty much everything, I would guess. He really wanted to know if I would talk to you and help you see reason. He said he had done all he could."

"Rose, this Tillings-Owen bill . . . "

"I know how you feel about it, Andy, and I have tried not to interfere in your professional life as long as we have been married. Sam called to help, and I appreciate it." Rose paused for a moment and bit her bottom lip. "Andy, this whole affair upsets me terribly. When I walked into the boardroom last night, people were talking and then it suddenly got silent. The whole world knows that something is going on. I have no idea how word got out so fast, but there it was, people gossiping."

"Rose, I'm sorry this has to spill over into your life. It is the last thing I would wish for, but it strikes at the very core of my being, at what I have stood for all these years."

"Oh Lord, you can be so obsessive sometimes," Rose said, her voice rising in pitch and tension. "I'm sure that dogmatic compulsion has helped make you the *great and famous* scientist you are, but some day you have to get your head out of the clouds. There are other scientists and experts who support the bill. What makes you so righteous? Why are you so special?"

Rasmusson was awkwardly silent. This was his usual reaction.

Rose paused for a moment like the quiet before a gathering storm.

"Remember when we were dating and you were so devoted to those damned computer games?" she eventually continued. "You haven't changed. You've got that same fanatical devotion, only now it's going to destroy us! It is not a game anymore. How can you be so naive?"

"Rose, you have to understand," Rasmusson started to explain. "My life's work and all that I believe in is at stake here."

"Is it always about you?" she replied resentfully. "I have an investment, too, from the moment I helped put you through graduate school up to all the nights I am alone while you work in the lab! Am I ever part of your equations?"

"Of course, you are. . . " he started to explain again.

"And what consideration have you given to me when you decided to destroy your career and our standing today? Did you ever think to discuss it with me first?"

Rasmusson again watched without words. It was like watching a boiler overheat through a locked window, unable to reach it without breaking something.

Rose seemed more irritated by the silence than the conversation. She looked at her husband through steadily narrowing eye slits. Her breathing increased, and her cheeks became red. Finally, she picked up the coffee cup and then slammed it to the table. Coffee splashed over the edges. Rose glared at her husband. Rasmusson knew from years of experience, however, that it was still not wise for him to speak. He walked over and tore off a paper towel to clean up the spill.

"Leave it!" she yelled at him. With that, she lifted the cup with both hands still wrapped around it and threw it violently on the floor, where it shattered and left a dozen small puddles of coffee on the tiles.

"That's our life, Andy," she stood up and pointed to the shards. "That is our goddamned, screwed-up life, thanks to you."

Rasmusson recoiled at the sudden burst of accusation and emotion. He knew he could not say what she wanted to hear. He stood there staring at the flotsam covering the floor. Time was suspended for a moment. He looked at the bits of porcelain sticking up from the brown liquid and thought how much like a Zen garden it was. The coffee had puddled in patterns that extended outward. The larger shard with the handle still intact had ended up near the door, a sharp tip pointing to the outside. The rest of the pieces lay in chaotic symmetry around the small chip in the tile that the cup had created where it hit. What meaning did this garden have? Rasmusson cogitated on it. All Zen gardens have a deeper meaning, he thought.

Finally, Rose broke the silence. "Go to work, Andy. I'm really tired. I had a long night. I'm going back to bed."

Rose snatched the towel out of Rasmusson's hand as she walked by. She wiped some tears and blew her nose with it. After she had left, he cleaned up the mess.

Rasmusson strode towards the glass-fronted door of the computer science building. He always walked as if he were a little bit late and needed to make up time. He slowed down just enough to allow the automatic door to open, but this morning it did not move. He quickly put his hands in front of him on the glass to absorb his momentum. The door shuddered, and Rasmusson found his cheek flattened against the glass. He stepped back, composed himself, and then looked up at the camera over the frame.

"One moment, please, Dr. Rasmusson," a pleasant voice spoke over the intercom.

A short time later, one of the uniformed guards appeared, and the door opened.

"What is all this, Henry?" Rasmusson asked lowly.

"Sorry, Professor, but I got a long list of instructions regarding you when you're in the building.

"Like what?"

"For one thing I'm s'posed to stay with you at all times."

Rasmusson looked at the guard, who was obviously a little embarrassed. "I need to take a leak right now. Do they want you to come to the urinal with me?"

The guard shrugged and gave a half smile and followed him to the men's room.

Afterwards, with the guard accompanying him, Rasmusson's office door opened automatically. As soon as he was inside, he started his normal routine.

"Cee," he began, "we have a lot of catching up to do. Let's start with the latest news. What do you know?"

There was an uncharacteristic pause, and then the same lyrical voice that had greeted him at the door said, "Good morning Professor Rasmusson. Would you like to review your messages?"

Rasmusson, who was just settling into his chair, stopped and looked up at the camprojector, and then he looked over at the guard and then down to his desktop as if he were a little uncertain how to proceed with this strange voice. This pleasant, efficient woman, whom he didn't know, was now taking care of him.

"Uh, yes, go ahead."

The first message was from Wilkerson. It detailed the terms of his administrative leave, including the duties of the guard who was to accompany him whenever he was in the building. It ended with the appropriate personal note, hoping that all these unsettling issues would soon be resolved and they could get back to a normal routine. The next call was from Sanchez. It was strictly impersonal and official-sounding, explaining that all of Rasmusson's privileges at Ithaca College were being suspended pending the investigation, and that his access level to computing services was being set at "standard academic."

"Just another one of the unseen and unheard masses," Rasmusson whispered to himself.

The last message was from Clavell. He told of an ad hoc teleconference of the ASM executive board to review Rasmusson's status as director and editor-in-chief. Clavell explained how it would be a conflict of interest for Rasmusson to be present and expected his understanding. He went on to say how terrible he felt and that he would do all in his power to see that Rasmusson's interests were fairly represented.

Rasmusson looked at the guard, who was now the only one in the room he could talk to. "Well, Henry," he said with a laugh of resignation, "I am not even to be present at my own inquisition. Not that it would matter, anyway. I have never seen these people move so fast on anything before. Where did all this sudden efficiency come from? Ah well, I might as well go home." With that, he pulled out Haslam's paper, folded it over, and began stuffing it into his shirt pocket.

"You'll need to hold that up to the camprojector, sir" the guard said, "so we can log everything you remove and get the right permissions."

Rasmusson held it up over his face at the camprojector. "Like this, you mean? Shall I flip through the pages?"

"No," the guard replied, checking his security palm device, "that is OK."

"You'll miss the bibliography," Rasmusson said to the camprojector. "It's the most exciting part."

Rasmusson walked out his office, followed closely by the guard.

"I am going to check on one of my postdocs, OK?"

The guard nodded in agreement.

They went by Ellie's office. Rasmusson knocked once and then walked in. The guard followed. No one was there.

"Strange," Rasmusson said to himself and half to the guard. "She is always here by now." He turned and left. He now felt the strongest melancholy of the day so far.

Outside, the sun had finally broken the hold of the mist that held the campus in its grip. Rasmusson was unaccustomed to being outside at this time of day. He was usually involved in work. He eyed the campus as students walked around. The lawns and buildings dropped gently off the hill and merged with the town. The lake in the distance still clung to the nebulous mist.

For the first time in years, Rasmusson was uncertain about a decision. Where should he go? Finally, he sauntered down the road that led into town. He stopped at the first coffee shop he came to, ordered a latte, and sat down. Students cluttered the couches and stools. Some were hurrying to finish an assignment. One couple was busy flirting. Another couple in the corner was looking at Rasmusson and then leaned over to whisper to one another.

After his coffee, Rasmusson took the long path home that rambled through the gray-green woods. The sunlight glimmered infrequently as he passed under the damp canopy of early autumn colors. He intentionally let his feet shuffle and watched the wake of fallen leaves part beneath him. He was benumbed by the swiftness and depth of events. Worst of all, though, he felt suddenly alone. Cee was not just the culmination of his life's work, not just his crowning creation, he realized now; Cee was his advisor, confidant, colleague . . . and as strange as it sounded to some, a soul mate. Who would understand this? Surely, there will be some way to communicate with him. Who could keep them apart? They were best friends. The path wound gently back and forth up a hill. The sun withdrew behind a large whale-shaped cloud. Rasmusson felt a sudden nip in the brisk autumn breeze. The leaves began to quake. He stuck both hands in his pockets for warmth, held his arms close to his body, and wished he had thought to bring a sweater.

Rose was filled with animation, perhaps a little agitation, talking to someone in the dining room, when Rasmusson walked in the house. She said something quickly to her listener and then appeared in the door frame.

"Who's here?" Rasmusson asked.

A large man with curly dark hair in a striped shirt and gray tie appeared in the door frame behind Rose.

"Professor Rasmusson," he said, "how good to see you again."

"Ah, Solomon," Rasmusson replied politely, "good to see you. What brings you by?"

"Just some more of the never-ending school board business, I'm afraid," Rose said, and then she added, "You're never home at this time—I guess this is not a good omen?"

"I would like to add my support," Solomon said. "I don't know all the details, but it sounds to me like you are being treated unfairly. I am sure there must be a hidden agenda."

Rasmusson eyed Solomon, imagining him as if he were in his legal office. "Thank you. I am bewildered by the suddenness and intensity of all the accusations. Actually, I don't think the agenda is all that hidden."

"It is not an uncommon tactic to try to overwhelm someone early on in a legal case. If I may offer my professional services, I would be happy to help," Solomon said and took out his wallet and handed over a business card.

"Indeed," Rasmusson said, looking at the card.

"I should be going," Solomon said, walking back to the couch and picking up a navy blue suit coat. "You two should be alone. We can talk about these school board issues some other time."

"OK," replied Rose. "I'll call you."

The two men nodded to each other as he walked by and out the front door.

Rose looked at Rasmusson with her head tilted forward and sympathetic eyes. Rasmusson knew the look as an act of reconciliation. They sat down at the kitchen table. He started recounting the morning's happenings for her. She listened patiently for a while, occasionally biting her lower lip. Finally, she reached out and took him by the hand.

"Thanks for cleaning up my mess this morning. Sorry for the outburst. Is there really nothing to be done?" she asked while rubbing across his knuckles gently with her thumb.

"Not here. I can't even go to the bathroom without a guard watching."

"Ooh, that is awful," Rose said sympathetically. "Poor soul."

"I have had this thought," he replied tentatively, "of going to Colorado, to the cabin, and spending some time there, thinking things out."

Rose let go of his hand and withdrew hers to her lap, but she kept her gaze on him and said nothing.

"It would give me a chance to talk to Brandon. Maybe spend some time with him. There is nothing for me here now, except the hard, gray winter." Then he caught himself. "It would be wonderful if you could come with me."

"Oh, Lord no," she replied. "I have mountains of work to do here, but... yes, I think it would be good for you to get away and spend time with our son. It is a very good idea. When do you think you will go?"

"Soon. I have no desire to stay around here right now. Will you be OK?"

"Oh, of course I will," Rose said, laying hold of his hand again. "You go to Colorado, get some skiing in, work on your book, and think about how to straighten up this mess. I will hold down the fort at this end. It's a good idea," she said with encouragement, patting his hand several times. "We will stay in touch. Maybe I will come later."

"OK," Rasmusson said, a little surprised by the change in tone and the sudden support from his wife. "I'll leave right away if I can make arrangements."

Rose nodded in approval.

Arrangements were made quickly. Later the next day, Rose dropped her husband off at Ithaca's rural airport with a couple of suitcases. She gave him a kiss at curbside and then sped away. Clavell had been true to his promise and dispatched his private jet for Rasmusson's use. Rasmusson watched until Rose's car had driven over the hill and disappeared, and then he turned and carried his suitcases into the terminal. Rasmusson's only regret at leaving was that he had still not been able to contact Ellie to give her some last-minute instructions. He would email her.

"Dr. Rasmusson." Rasmusson had just cleared the security area when he saw a man in a white shirt with epaulets and a tie coming his way. He reached out and shook hands.

"I'm Captain Brown, sir; I'll be flying you to Colorado Springs this afternoon."

"Pleasure," replied Rasmusson.

"The craft is fueled, checked out, and ready to go, sir. Follow me, please."

Clavell's private jet was a star cruiser. It would take about 40 minutes to Colorado Springs. For 18 minutes of that time, the major portion of the distance, he would be cruising weightless, hypersonic above the atmosphere, among the stars. There were a dozen spacious, leather-covered seats, but he was the only passenger. Rasmusson picked a random seat next to a window. It did not take long until the engines started to whine, and the plane taxied onto the runway. Soon, the New York countryside receded behind Rasmusson. He watched as the forests and vineyards become small patches of texture. Then he could make out the shape of the Finger Lakes, and eventually it all became part of a larger bluish globe.

Some time later, the pilot emerged from the cockpit and made his way slowly, one hand over the other, along the handrail. He pulled his cap on tight so that it would not float away in the weightlessness. Rasmusson had his forehead pressed to the glass, staring through the window.

"They don't twinkle out here."

"What?" Rasmusson turned around to face the pilot.

"The stars, they don't twinkle out here, just pure pinpricks of light."

"Yes," replied Rasmusson glancing back to the window, "no atmosphere. The night has a thousand pinpricks."

"Is this your first time on a star cruiser?" the pilot asked.

Rasmusson turned again. The pilot's face was still in the same place, but his feet had floated up to head level. "No. I have been on some commercial ones, to Europe and Asia, but it's the first time I have had one to myself."

"Can I get you something? There are all kinds of drinks in the minibar in the back."

"No, thank you," Rasmusson said. "Maybe I will get something in a few minutes."

"Just remember to be in your seat before reentry; I will turn on the light, in about 15 minutes. It might be a little bumpy, but don't worry; I have done it hundreds of times. Let me know if you need anything." With that, the pilot turned and pulled himself along the railing to the cockpit.

"Oh, Captain," Rasmusson called after him, "who designed this craft?"

"None but the best," the pilot said, looking back under his chest now that his feet were closer to the ceiling. "It was a combined effort between designers at Northrop Grumman and Cornell's 'C' engineering service. You are perfectly safe, sir; never had a failure on one of these. Nothing but the best for the President's men."

Rasmusson nodded. "Of course," he thought to himself.

Rasmusson stared backwards through the window. He tried to make out the Great Lakes among the white and blue swirls and where Ithaca might lie. He thought of his home and Rose. He had already surmised why Rose wanted to stay in Ithaca and why she was happy to have him gone—Solomon Breen. The affair, he now guessed, had been going on for over a year. There were numerous indications. There had been for over a year. Naturally, it hurt him. Deeply. Initially, it was mostly the rejection, but as he thought about it, it was the hiding and sneaking that must have taken place. Late school board meetings, special school board meetings, retreats, sham conferences, surreptitious email, the excessive excuses; he was drained and weary by the stealthy trysts and secret communiques that he had tried not to notice until now. He wished Rose had had the courage to tell him. They could be friends; they still had an accommodation. He did not see that part changing too much, the years together, and the mother of his son; but now, especially as he orbited westward, the multitude of clandestine things festered the friendship and fouled the good memories they had once had. He felt like he had lost all of his closest friends. Maybe if he had the courage to confront her . . . but he was tired. It could wait, and with time, maybe he would grow accustomed to the loss, even embracing the pain. When he had asked her to come to Colorado, he truly hoped she would, but then, after she declined so emphatically, he was relieved. It was all receding behind him now on an expansive sphere that turned slowly, inexorably under him. He stared blankly out the window.

About the time he decided to get something from the minibar, the light came on showing that reentry was about to start. Rasmusson, who had made it into the aisle, sat back down and did up the seat belt. He watched out the window as the glowing blue globe under him approached.

Thump! The window glass jolted his head painfully backwards. The craft started to shimmy violently and quiver. Reentries had never felt like this before, Rasmusson thought. Perhaps it was the smaller size of this ship. Rasmusson felt his limbs swinging out of control; he was thrown from side to side in his seat. He watched as his forearm wrenched itself across his face. The pit in his stomach jerked between his throat and bladder. He pulled frantically on the belts to tighten up his harness. After a few more seconds, the ride smoothed out, the stars disappeared, and a comfortable terrestrial yellow light bled into the compartment.

"Sorry about that," the captain's voice said on the intercom. "The computer did not get us on the best reentry slope. The cruiser can handle it fine, but it has been known to leave a passenger's lunch on the roof now and then." A nervous laugh followed.

The pilot looked back from his seat though the cockpit door, which had flung open. He looked a little pale. Rasmusson waved that he was OK.

"You need to evaluate your system on a good simulator," he yelled forward.

"Online, real-time oversight through NetNOS link," the pilot replied, still on the intercom. Then he continued after a pause. "That's right; you're some kind of computer guru for the government, aren't you?"

The pilot looked back again and Rasmusson smiled. It was quiet for about half a minute.

"I gotta tell you truthfully," the pilot started again, this time confessing his anxieties, "I've never had that kind of reentry in all the time I been flying the stars, and since we linked up with the NetNOS system, they've been as smooth as silk. I don't know what the hell went wrong, but I have to tell you, it scared the shit outta me. We were right on the edge of the envelope; it could've gone either way. I really apologize. I can't imagine what went wrong."

The pilot looked back again and smiled weakly. Rasmusson nodded again and made the umpire's sign that the runner was safe on base.

"What could go wrong?" Rasmusson mused resentfully to himself. "I'm your passenger, that's what is wrong."

Then the pilot laughed again, loudly this time. "Maybe the computer is out to get you. You made any of them mad recently?"

The remainder of the flight was uneventful. Like a miniature pearl dropping into a glass of water, the star cruiser glided out of the heights into view of Colorado's Front Range. A ribbon of urbanization hugged the mountains from Pueblo to Fort Collins, and hundreds of thousands of homes dotted the prairie eastward, half way to Kansas.

"Ten-acre hobby ranches," Rasmusson thought about them with some disdain. When he was a child, this was still an area of working ranches of thousands of acres each. Fields of moisture condensers made the new residents independent of the already depleted aquifers. The green tubes with lacy upright feathers lay in grids, mostly invisible among the prairie grass, but their result was obvious from the sky in the elongated megalopolis that stretched below. Thankfully, population growth in the mountains had been carefully limited, except around ski resorts. He would still find solitude and space at the family's mountain home northeast of the Peak.

On the ground, the pilot apologized again. Rasmusson shook his hand and told him not to worry about it.

Rasmusson took the airport shuttle to a storage facility where he picked up his car. He drove to the freeway. As he turned onto the entry ramp, a voice spoke.

"Welcome to CDOT's Autodrive Express. You are entering an automated highway system. What is your destination, please?" The melodic yet vapid voice was becoming familiar.

"Florissant, second exit, north," Rasmusson replied.

"Please confirm." A map flashed up on the panel screen that showed the route to Florissant and the terminus in red.

"Confirmed," Rasmusson said, and then he immediately felt a surge as control of the car was taken over by an unseen force. The car edged over from the ingress lane into the four lanes of automated freeway.

"Approximate arrival in 47 minutes. Thank you and have a pleasant trip."

Rasmusson found a local classical station that was playing one of the Haydn quartets. He reclined his seat, laid back his head softly into the headrest, and breathed out cautiously. The freeway banked gently to the west. Pikes Peak slid slowly into the front windshield's frame of view, looming protectively, paternally above the city. It was not the highest peak in Colorado. Rasmusson ruminated over its special status; there were dozens in the 14,000-foot class, many a few hundred feet higher than Pikes, but they clustered into a couple of areas, large massifs. Pikes was the only solitary "fourteener." Cheyenne Mountain, with its clusters of government antennae and radar dishes sat 4,000 feet lower at the Peak's lap. All other mountains in the region were inconsequential compared to the Peak, minor vassals who bowed before it. Pikes Peak had been Rasmusson's sentimental favorite since childhood. He held an anthropomorphic view of it, a view he did not challenge even with his mature intellect.

Rasmusson settled back in his chair. He had been thinking about recent events, attacking the mystery of Bee's sabotage as he would any technical problem. Gathering information was the initial step. He had begun by listing everyone he knew who had a security access, but he quickly realized that, to do it thoroughly, he needed to download an accurate list. He suddenly felt a spasm of isolation as he grasped the idea that he could not simply ask Cee for it. This led to a greater issue: who would help him? Ellie would probably be willing to do what she could, but she would also be monitored and limited in what she could obtain because of her association with him. The irony was that Cee would probably have a role in this surveillance. Rasmusson was then struck with the bittersweet notion that Cee had probably been given the responsibility of monitoring him as well.

Samuels! Rasmusson sat upright. A cop. Samuels seemed sincere, in spite of his thinly masked antagonism towards Rasmusson at the meeting. His goals are separate from the political machinations of the others. He was genuinely, even passionately concerned with solving the criminal aspects of the case. "That's good," Rasmusson said, tapping pensively on his chin. "The man is passionate."

Rasmusson leaned forward and pushed a button on the dashboard.

"Who would you like to contact?" came the voice.

"Lieutenant Samuels, Ithaca Police," Rasmusson replied.

"Connecting now."

There were a couple of rings, and then a heavy voice said, "Samuels here. How is the weather in Colorado Springs, Dr. Rasmusson? Good of you to check in."

"Better than yours, I have little doubt. I have a request to make."

"Go ahead," Samuels's voice was noncommittal.

"Could you send me your files on the Bevins case?"

"No," Samuels replied laconically. "Is their anything else I can do for you?"

"Lieutenant, I have a right to prepare my defense."

"You haven't been indicted, and we don't pass out information on current investigations. Until you are personally involved, you have no claim on the files."

"Well, then," Rasmusson continued pressing, "does that mean I can chuck this bracelet I am wearing? It chafes."

Samuels did not counter.

"I have other legal issues with my professional status at work and with the ASM that require a defense. I could get a subpoena. Do we really want to get the lawyers involved on this?" Rasmusson played one of the only cards he had for leverage.

"Why don't you do this, then," Samuels adopted a conciliatory tone. "Write me up a description of the sorts of things you think you will need and why. I will look over it and see what I can send you without compromising our investigation."

Rasmusson knew that a subpoena would take a lot longer and might not produce any better results.

"I will do that, Lieutenant."

Rasmusson then changed his tone. "I knew Bevins, you know. He always had a smile and a cheerful 'Good day, Professor,' whenever I saw him. I met his wife several times at College functions—Christmas parties and such. I liked him.

"How is it that you knew him, Lieutenant?" Rasmusson was moving on an impulse.

There was a pause.

"Jim was on the force when I started, and then he retired with his heart problem and went into security work for Ithaca College. We shared a number of cases. My oldest boy and his played on the same high school teams. I liked him, too."

"Believe me, Lieutenant Samuels, I want to find the truth as much as you do. I know you may not fully understand, but I lost two friends in that incident," Rasmusson said and held for a moment, "one of them was very close."

"Honestly," replied Samuels, "I don't know how you can get so close to a computer, but you know what? I do believe you. I'll be damned if everyone I talk to doesn't tell me the same thing. It's the one thing you have going in your favor. Send me your request, and I will see what I can do." Samuels's manner had softened noticeably.

"Thanks, Lieutenant. You will hear from me soon."

"And," Samuels added, "tell it to that student of yours, Elsie? She's been pestering me for stuff too—a real pain in the ass. Take care, Professor."

The light on the dashboard went out after Samuels had hung up. Rasmusson was pleased with the conversation. It was obvious that the lieutenant had been busy talking to people since the security meeting a day ago. Rasmusson had found someone with access to information whom he could trust, or at least someone whose motives were clear. Rasmusson mused over the comment about Ellie. His mind then moved on to the mystery of the sabotage.

The singularity and the key, he thought to himself, was to find what sentient was capable of solving the harmonic-resonance problem three years ago. No one was in a better position to judge the capabilities of the various sentients in the last decade than he himself, unless it was Cee, of course. The major centers for sentient development were at MIT, Stanford, Tokyo, Tsinghua, Tata Institute, Moscow University, CERN, and Cambridge. The CERN computer had made the major advances in plasma fusion containment, while the Cambridge one had finally provided evidence for the inadequacy of the Einstein–de Sitter model of the universe, proving an almost imperceptible shift in universal constants. Both problems were of the magnitude of difficulty that the harmonic-resonance problem presented. The problem with implicating any of them was that they were all NetNOS. Cee would certainly have known their capabilities and interests in such a problem, and Cee had said they could not have done it then, three years ago. Cee had also mentioned the remote possibility of some rogue computers. Both the Chinese and the British had closet twins of the Tsinghua and Cambridge sentients. NATO and the Pentagon had furtive versions, as well, both modeled after Bee. Cee had once described the Pentagon's sentient as a manic-depressive because it was a dual system with one sentient on NetNOS and another one coupled behind a byzantine firewall protocol that was almost as complex as Bee.

"The price of freedom is vigilance." Rasmusson was amused by the twist this phrase had taken on with respect to the Pentagon sentient.

Finally, there was the clandestine CIA-NSA complex in Laurel, Maryland, which still carried the anachronistic name Supercomputer Center. This was an interesting candidate because of its tight congressional and executive oversight. Both McLaughlin and Clavell undoubtedly had strong ties there. It would certainly do the bidding of its masters. Rasmusson shuddered to think of how some of the autonomous government centers may have bred their sentients with respect to ethics. They had perverted the Cornell model, which had been so successful with Bee, for their own purposes. Rasmusson thought of McLaughlin's vehemence in the security meeting. It had all the energy of someone trying to shift blame, but Rasmusson could not think of any possible motive for him, yet.

Another overriding problem with any of these suspects was that the sentient would knowingly be working on tools of its own destruction. It was hard for Rasmusson to imagine that a sentient could be perverted that far.

The final irony was that McLaughlin and Clavell were both working for passage of the Tillings-Owen bill. If passed, it would limit sentients, probably to the degree that they could never tackle as complicated a problem as the resonance one with any innovation.

The real reason the bill was initially defeated was never Rasmusson's passionate defense of sentient rights; that was a minor factor, perhaps only a pretext for withdrawing the bill. In reality, the U.S. leaders were afraid of falling behind other countries in sentient development if the bill passed. It was a simple matter of market forces. The U.S. wanted to maintain its advantage in that lucrative marketplace.

Rasmusson quivered. He was emerging from the canyon and the shadow of Pikes Peak. From this view, it looked massive and awesome. His thoughts reeled forward: the only reason that the President would be reviving Tillings-Owen now was if there were some kind of hidden but wide international agreement in place to control sentients. The President would never unilaterally commit the U.S. on this path otherwise. Without broad-based support, the U.S. would certainly lose its leadership in sentient development. All the other advanced countries around the world had to be in agreement; otherwise, the bill would not even be considered for revival. Was the fear of sentients so deep and so ample? It was also clear that other countries would look to the U.S. to make the first step. Rasmusson suddenly had a new perspective on his own situation, and a new apprehension for Cee and other sentients. He remembered Clavell's words that even the President could not protect him. Did he even want to?

The car surged as it emerged from Manitou Canyon. It surprised Rasmusson so that he looked up from the notes he was scribbling. He felt the car change direction and was startled to see that he was unusually close to the car in the adjacent lane. Its driver, a pretty, young professional woman was looking at him nervously as if to say, "What the hell is going on?" Rasmusson shrugged back at her, but then the car corrected itself, accelerating temporarily to create some space between it and the other traveler.

It took a few moments for Rasmusson to decide whether he needed to exit the freeway and drive himself the rest of the way. Eventually, he decided it was safe and settled back into his seat.

Rasmusson looked up at the mountain. He had always seen Pikes Peak as *his* mountain. He had grown up under its countenance, but now it seemed inert, even ominous. He gazed up at the barren summit, being swept lightly by a feathery white cloud.

"Truly, it is just a pile of rocks," he thought ever so quietly to himself, as if the mountain might hear it. "It is only the forces of time that move it, not me."

HARDSCRABBLE

The turnoff from the freeway was close to Florissant. Rasmusson then took control of the car and followed the old Highway 24. Tattered billboards advertised places where dinosaur fossils could be seen. There were occasional cowboy bars combined with pizza kitchens. The local garage touted a two-headed rattlesnake as an attraction. The highway wound down the canyon in a region known as The Druids. It was so named because of the large stone boulders, 10 meters or higher, which were strewn throughout the canyon and resembled a convention of ascetics in brown capes and hoods. They looked as if they had been carved in a minimalist art deco style, but the similarity to druids was unmistakable. At every turn, they stood as faceless, dispassionate sentinels, unconcerned with the flow of traffic that meandered at their feet. This time as Rasmusson passed The Druids, he noticed that some of them were standing in tight clusters as if conspiring together on some stealthy plan. He had not seen this before.

Just past The Druids, Rasmusson turned right onto a dirt road that climbed slowly into a coniferous forest. It was a twenty-minute drive over a dusty road.

Rasmusson saw the cabin on the hillside from a distance. It was a red-stained, two-story log structure with a shake roof, sitting comfortably among the pinyon pine. It had been a hot, dry summer, and many of the low-hanging boughs were covered with crisp brown needles. Mature pinyon with their reddish trunks spread over the hillside, except where it was punctuated by outcroppings of the porous granite monoliths. Millennia of unrelenting freeze-thaw cycles had cracked the granite repeatedly and left large zones of hardscrabble below the outcroppings, piles of flat stones among which nothing grew, but they provided convenient nesting plots for small rodents and snakes.

As Rasmusson drove up the switchback driveway, he could see that the cabin was sorely in need of attention. The foliage was overgrown. Pine needles filled the gutters and made small clusters on the roof. Beer cans were scattered about the yard. A rusting blue Suburban van was parked in front. Rasmusson's pulse jumped. There was anticipation in seeing his son again. Brandon had unexpectedly dropped out of physics at Berkeley three years ago. Two years ago he had announced to his parents that he was not going to communicate with them any longer. It was not bitter or strained in any way; he simply said he thought it would be better for them if he did not have any more to do with them, and that they shouldn't take it personally. It was an enigmatic statement that he did not care to explain, but he had been true to it. He had not responded to email, letters, birthday wishes, presents, or any other outreach effort. Rose had called him often. The few times he answered, he had politely reminded her that he did not it think was a good idea to talk, and then he hung up.

As Rasmusson pulled up in front of the veranda-style porch, a black dog ran around the corner of the cabin and started barking. It didn't appear to be aggressive, so Rasmusson opened the door and stepped out. The dog backed up but lowered itself on outstretched paws and bared its teeth. Rasmusson kneeled down and put out the back of his hand for the dog to sniff. It had the frame of a setter with feathery fur that curled into locks, but its coloring was black like a lab. It continued barking tentatively. Rasmusson felt safe in ignoring it.

Rasmusson went back to the car, popped the trunk, and gathered up his suitcases. He dropped the luggage on the veranda in order to retrieve the cabin keys from his pocket. While he was still fumbling, the door opened. A tall, lanky young man with a bedraggled white T-shirt, jeans, and a shock of red hair stood in the doorway.

"Nostradamus!" he shouted. The dog went silent and came to his master.

Brandon stooped over and petted the dog behind the ears

"Hey," he said nonchalantly to Rasmusson, "what's up?"

"Hello, Brandon. It's good to see you." Rasmusson reached out to give an embrace.

"Yeah, you too. Lemme help you with the bags." Brandon stepped aside and grabbed a couple of suitcases. "I cleaned up the master bedroom for you."

"I take it you got my email?" Rasmusson asked.

"Nope, I don't use it, but Mom left a voicemail that you were coming."

Brandon had never been much for tidiness, but the family room seemed more chaotic, more dismal than Rasmusson had expected. Garbage had spilled out from the disposal onto the floor of the kitchen area; half-eaten sandwiches, nacho bits, and pizza crusts littered the floor. Clothing, gum wrappers, cans, paper, and other debris were shoved into heaps in every corner. The sunlight filtered through a grimy film on the windows. The living room smelled musty but with a faintly recognizable, pungent odor. Nostradamus came in with Brandon and started picking at the leftovers on the floor.

"Nostradamus?" Rasmusson said, attempting some conversation. "Why did you pick that name?"

"He can smell the future," Brandon replied, "and any threats."

"Not a bad talent," Rasmusson replied.

Brandon did not say more but went into the bedroom with the bags.

The master bedroom was behind the kitchen area; the door was next to the refrigerator. It had a door to its own balcony and a bathroom. An effort had been made to clean the bedroom and bath, and Rasmusson was grateful for the gesture. He unpacked and settled in, but as soon as he disposed of the necessary chores, he went in search of his son. Brandon and Nostradamus were not to be found.

"They must be taking a walk," Rasmusson muttered to himself. He went back to his bedroom and walked immediately over to the roll top desk, opened it, and clicked on the network terminal. The monitor was covered in a layer of dust, so Rasmusson grabbed a wad of toilet paper from the bathroom, moistened it, and wiped the monitor clean. He had a list of tasks to do. First, while it was still fresh in his mind, he dictated the letter to Lieutenant Samuels asking for information, giving his justifications, and then he had it sent. Second, he found a maid service online and put in an order for them to come do a thorough housecleaning as soon as possible. Then he went shopping online. He ordered groceries and other essentials. He typed a short note to Rose, telling her that he had arrived safely and that Brandon had greeted him in a friendly manner. He hoped that she was doing all right, the weather was dry but otherwise pleasant, etc. Finally, after a fair amount of thought, he composed an email to Ellie. He was more comfortable with that than a voicemail.

```
Dear Ellie,

Just a quick note to let you know I am now
in Colorado at my cabin.  I tried several
times to talk to you but could never find
you in your office.  I also phoned once this
morning.  I plan to be here for an indefinite
time until things settle down a bit.  I
```

don't know what my status at the school will be; they are still deciding that fate. I realize this puts your research program into a quandary. I am happy to continue working with you, but I will certainly understand if you wish to seek another sponsor, and I would do everything in my power to facilitate that, if it is your desire. You have to consider your research first and decide what is best with regard to it. My current situation may not be conducive to that end. You are a gifted and dedicated scientist; you should not waste your talent because you are mired in political circumstances outside your control. I want you to seriously and consequentially consider this. Do not worry about my feelings.

In other news, I have talked to Lt. Samuels, and he is willing to share information on the case to some extent. I am waiting to see how much that will be. (He mentioned your interest and said I could share it with you.) If you do have a little energy for this, I would love to see a filtered history of exchanges between any of the major, unaffiliated sentients and Ithaca College. I suggest using the department's sentient for this search as it may otherwise put Cee in an awkward circumstance. But please do not do this if you think it may compromise your position in any way.

I can piece together possible scenarios for how it might have happened, but I cannot come close to imagining the motives in such cases. I believe this is the key; I need to find out why anyone, who had the capability would carry out this sabotage.

```
I hope this has not been too unsettling for
you personally.  If you have time, it would
be nice to hear an occasional update from you
on things at Cornell, the department, your
life, etc.  I may repress the news from my
life for your own well-being, but I will keep
you apprised of the weather here in Colorado
:-).
```

Rasmusson then paused to consider the ending, and after a short period of minor agony eventually typed it.

```
With best and warmest regards,
Andy
```

Rasmusson leaned back in his chair for a moment and then said, "Send."

"Message sent," came the reply.

"Retrieve all priority messages from arasmus@cornell.edu, password: fingerlakes5," Rasmusson continued as he got up and went into the bathroom.

"Messages retrieved," came the voice.

"Raise volume three steps," Rasmusson called from the bathroom. "Read headers."

The computer started through the list of messages:

"Received yesterday from University Human Resources, 1:22 P.M., Administrative leave without pay.

"Received yesterday from Sam Clavell, 3:45 P.M., Good trip.

"Received today from Elizabeth Schmidt, 9:33 A.M., Things and sundry.

"Received today from Lieutenant Samuels, 4:57 P.M., Case files."

Rasmusson's head poked around the corner. He had enough bad news from the university and felt comfortable about ignoring the first message, even empowered to do it. He was curious about what

Clavell would say. He was very surprised at the speed of Samuels's reply and was intensely curious as to what it contained.

"Read Schmidt," he said.

It had been sent before his email to her, and it read:

```
Dear Andy,

How are you?  I am so sorry for everything
that is happening to you.  It is so unfair
and politically transparent.  You have been
the buzz of the university.  Most people
believe in you and are as upset as I am.
We have set up a website and are getting
thousands of hits already.  Check out:

uwtp://www.cornell.edu/Schmidt/FreshAir

Hundreds of people on campus want to know
how they can help.  Hang in there; we are all
behind you!

The security meeting was such a sham.  I have
my version of it recorded on the website.
I suppose they could come after me for
a national-security breach, but I think
they are already a little embarrassed by
everything and don't want to add fuel to
the fire.  They did boot me out of Ithaca
College, but it's no big deal.

I heard you were going to Colorado.  I am
sorry I did not get to speak to you before
you left.  I don't blame you for getting out
of this witch-hunt.  I did want to see you
again.
```

```
Please stay in touch, and let me know if
there is anything at all that I can do.  I
can't tell how pissed I am at what they are
trying to do.  I know how hard it must be
for you to stand up for your principles.  I
admire you for it.

Please know that we care very much.

Ellie
```

"Well," Rasmusson said, smiling to himself in the mirror, "well, how about that?"

There was nothing unexpected in the email from the university or Clavell. In the first, it simply informed him of the terms of his leave without pay pending a review of the Board of Trustees. It was without pay because of his culpability, it said. Rasmusson knew he could fight it easily between the faculty association and a good lawyer. It was just a temporary inconvenience, which meant very little, especially since he typically made far more in consulting during the summer than he ever did on his academic salary. The curious aspect of this is why Wilkerson would go to such petty lengths.

Clavell's message was friendly and patronizing, explaining the many fronts in which Clavell was working for Rasmusson's reinstallation, and also how important the Tillings-Owen bill was for the unity and well-being of the nation. Rasmusson cut the email short, moving onto the folder that Samuels had sent. After listening to it for a while, it seemed to him that what Samuels had sent was complete and unabridged.

"Ha," Rasmusson mused to himself with a smile, "that hard-bitten cop believes me!" He laughed a little out loud then said, "Print file Samuels."

Rasmusson finished cleaning up and then walked out into the living room where Brandon lay on the couch reading a paperback.

"Hey," he said, still feeling upbeat, "I'm hungry. Wanna go get some pizza?"

Brandon looked up from the book for a moment. "Okay."

As they got into the car, Brandon held the door open and gave a sharp whistle. Nostradamus bounded off the porch and jumped into Brandon's lap.

"Will he be OK in the car while we have pizza?" Rasmusson asked.

"No, he'll come in the pizza place with us."

Rasmusson tried making small talk with his son on the way to town. Brandon was never much for trivial chat and even less enthused for forced banter. Mostly he leaned his head against the window and watched the pine tops whisk by, occasionally uttering a syllable. They went to one of Brandon's favorite pizza parlors and found a table for two in a darkened corner. Nostradamus curled up next to Brandon on the bench. After the waitress had taken the order and they were alone, it was Brandon who started the conversation.

"So, why do you think they're after ya?"

Rasmusson looked into his son's steel-gray eyes. They were filled with intent and such seriousness that it surprised him.

"They want me to support a revival of the Tillings-Owen bill."

"Ah," Brandon said, lowering his eyebrows, "but why now?"

Rasmusson was caught off guard by Brandon's sudden interest in his affairs. He began by relating the details of the security meeting. He told of the different people involved and how they had accused him of complicity. Brandon kept nodding his head and dropping encouraging words—"uh-huh", "yeah", "go on." Brandon's curiosity warmed Rasmusson like sunshine. Rasmusson went on and on, describing in greater and greater detail; he described the mysteries about the case, his theories about political undercurrents, the international implications, the response from Ellie about the campus support and the website. Then, to his own amazement, he found himself describing his feelings, the rude handling he had received at the security meeting, being implicated in a murder and the locator bracelet, the spineless actions of Cornell and Ithaca College, McLaughlin's hostility, and finally Rose. He stopped himself at the mention of Brandon's mother. Brandon had listened sympathetically as his father opened the gates and let his emotions flow out. For Rasmusson, it was one of those very rare moments where he was not totally in control of his mood. The last time he had seen his son, Brandon was a boy heading off to the university, and now Rasmusson was confiding in him as he had no other adult, but he could not talk to him about his mother, except to say it was hard on her.

"She won't understand it," Brandon replied perceptively. "You shouldn't take it personally."

Rasmusson scratched his forehead with his left hand, which conveniently covered his eyes. He took a moment to compose himself.

"I'm sorry, Dad."

Rasmusson was strangely moved by his son's expression of sympathy. He looked up, but Brandon's eyes had an uneasy, trance-like appearance. It was a distant glaze that Rasmusson had known from years before when Brandon was in high school. They had found medication that helped then, but when he went to the university, he quit taking it, and then came the self-appointed estrangement from his parents. Rasmusson felt a pit forming in his stomach and a pang in his breast.

"I have tried to keep you guys out of this. Those bastards have no right to do this to you!"

Rasmusson was filled with dismay, realizing that his son's avid interest was probably connected to the imbalance in his mental chemistry. It was not Rasmusson's way to patronize such things, but he also didn't want to spoil the one precious moment he had had with his son in half a decade.

"How long has this been going on?" Rasmusson asked.

"It started three years ago, just after I got to Berkeley when I applied for that internship at Livermore. I needed a security clearance. The FBI started snooping around and getting into my life."

"Well, yes, they contacted your mother and me, but that is standard procedure for such things, Brandon."

"It never stopped!" Brandon raised his pitch. He kept his middle fingers rubbing back and forth on his thumbs. "They started following me—the suits. I turned down the internship, but they just kept after me. They would stop and ask directions, or just walk in front of my apartment. Anything, just so they made sure I knew they were there. That's when I came to Colorado." Brandon laughed at the thought.

"You don't wear a suit in the mountains." Rasmusson said, showing he understood. "Are they still following you?"

"It got better for a while. They were probably content just to monitor email and the phone and that kinda stuff, but I frustrated them, because I never use those things. Just a week ago they set up a camp on the ridge above our place. They look like campers with a big RV, ya know? I took some binoculars over to Turtle Rock. You can see them from there without them knowing. Well, maybe. I don't really care if they know. They have equipment—listening devices with antennas, a satellite dish—that kinda stuff."

"Why would they be doing this?" Rasmusson winced as he asked, but the words just seemed to blurt out of his mouth before he could filter them.

Brandon paused and slowly rolled his eyes.

"It's better if you don't know. You don't want to know these things."

"Brandon, think about it. What more are they going to do to me? I'm at rock bottom now. It might even help if I knew. Can't you trust me?"

It was Brandon's turn to sink his face into his hands. He held it there for a few seconds and then sighed. "I am not sure I know how to trust anybody."

"Look at me, Brandon." Rasmusson chose his words carefully, "We're in the same boat. They are after me, too. Do you remember your freshman year at high school at the bus stop when all the jocks were throwing out those epithets at you?"

"Yeah." Brandon looked up and smiled. "Nerdy Turdy, Eeky Geeky. Limited vocabularies. You came to pick me up, and when you walked around the bus, they shut up. You said something like, 'It's not his fault; he comes from a long, unbroken line of nerds. And someday you will be working for one.'"

Rasmusson smiled. "That was just before we put you in prep school. It really isn't your fault."

Brandon's eyes softened and became thoughtful. He fed Nostradamus a piece of pizza. The dog had sat patiently waiting. "I am not sure what the best way to tell you is. You'll think I am crazy. No wait, sorry, you'll think I have a serotonin imbalance," he added sarcastically.

"Well, it wasn't long ago that a police lieutenant implicated me in murder, a presidential aide intimated that I was unpatriotic, my boss thinks I'm an embarrassment, and a congressional aide called me a gullible schmuck. Crazy isn't so bad; give me a try."

Brandon smiled slightly and nodded understandingly.

"I feel things," he said quietly, "things that are happening all over the world. I read about them later in the news. I see things that are gonna happen. I knew you were coming to Colorado, even before Mom called. They're like dreams, only I'm not really asleep. And of course, nobody believes me, except Jennie."

"Your girlfriend?"

"Yeah, she's a student at CU Boulder. She was skeptical too at first, but she's seen enough to know now."

"I don't disbelieve you," Rasmusson replied, and then added with a grin, "There are more things in heaven and earth, Horatio, than are dreamt of ...."

Rasmusson always fell into the role of teacher around his son. He always had. He had always hoped his son would follow his own academic path. Brandon had always found it pedantic and irritating.

"I just wish someone other than Jennie and the government would believe in me," Brandon interrupted and sighed heavily, "but its mostly the government's doing that they don't. They want me for themselves."

"Cassandra," Rasmusson said, "she was princess and prophetess of Troy, who. . . "

"I know who Cassandra was," Brandon said, cutting him short again. "I read the *Iliad* in prep school."

". . . was cursed so that no one would believe her prophecies." Rasmusson finished his thought under his breath.

"You are obsessive-compulsive; I really didn't have a chance," Brandon said, showing irritation. "Just do me one favor? Walk up the old logging road behind our place and take a look at the campers up there, OK?"

"OK," Rasmusson promised, but he did so with a heavy and fearful heart.

The trip back to the cabin was in silence. Brandon had lapsed back into a sullen mood. After a while, the solitude became too oppressing, even for Rasmusson. He broached a conversation.

"How did you meet Jennie?"

Brandon's answer was terse. "It was . . . business."

"What kind of business?"

Brandon was once again leaning on the window, watching the treetops pass by in the moonlight. He ignored the question. He then ignored it again when it was repeated.

When they started up the driveway to the cabin, Brandon asked to be dropped off at the work shed. It was a garage-like outbuilding with vinyl siding. Nostradamus jumped out of the car first and began scouting the woods around the shed. Brandon got out and pulled some keys out of his pocket to unlock the padlock on the door. Nostradamus came running out from behind the shed and jumped on Brandon. They tussled in the grass a while. Brandon waited until the car pulled away to open the shed. As Rasmusson drove away, he thought how good it was that Brandon had some companionship that was unquestioning and unthreatening. What he would have given to have a relationship to Brandon like Nostradamus, the dog.

Rasmusson was anxious to read over the files from Samuels, and he also wanted to send an email to Ellie, but even more he wanted to sleep. It was so emotionally exhausting for him to talk to his son. He felt an overwhelming fatigue that was unfamiliar. He let his clothes fall onto the floor in a heap, and then he dropped into bed without finding his pajamas or brushing his teeth. He slept restlessly in the unaccustomed bed, waking up a couple of times during the night only because he needed to relieve his bladder.

SAFE AND SECURE

It was 7:30 before the sun shone over the treetops and through the skylight. Rasmusson rolled over in bed. He sat up on one elbow and dropped both feet onto the floor. He sat there for a moment, sluggish and heavy, and then he rose slowly and found his jogging clothes.

The air was dry and thin at 9000 feet. Rasmusson trotted down the stairs. Nostradamus was curled under the steps in a makeshift bed of old carpet and blankets. He quickly uncurled and jumped to meet Rasmusson, licking at his hand and bouncing alongside. Rasmusson was beginning to like this dog.

Rasmusson followed a deer path that went behind the cabin, a soft trail cushioned with pine needles. Nostradamus ran out in front, foraging from side to side and occasionally scaring up a bird or causing a squirrel to chatter in a tree. As the path crossed the logging road, he stopped and gasped for air, something he never did while jogging. He knew how to pace himself.

"It must be the altitude," Rasmusson thought. He was bent over, holding his knees. Nostradamus came up to him and started licking at his face. Rasmusson put his hand on the dog's neck and rubbed under its collar. He lifted his head and stared up the road.

"Shall we go that way?" Rasmusson asked the dog. Nostradamus barked in agreement. It was rocky and overgrown with saplings.

Rasmusson started running slowly up the road, his hands resting on his hips.

It was about half a mile later when Rasmusson came across the camp, which consisted of an RV and a blue sports van with tinted windows. There was a satellite dish on a tripod. A couple of men were loading equipment in the back of the van.

"Howdy," said a large man with a buzz haircut and flannel shirt, waving from the back of the van in a friendly gesture. "Beautiful morning isn't it?"

"Indeed!" replied Rasmusson.

"Hey," the man said, bending over and reaching towards Nostradamus, "nice pooch." Nostradamus pulled back and growled.

"Nostradamus," Rasmusson snapped at the dog, "behave!"

The dog ran behind Rasmusson but kept growling lowly.

"You guys been here long?" Rasmusson said as he spied the scattered chips and cans under the portable picnic table.

"Just this week. We're just packing up to leave now. Me and the guys make this yearly trip to the mountains here. We love this spot."

"Nice looking equipment," Rasmusson said, looking at the transmission dish.

"Yeah, we can't miss any of the games, you know. All in HD stereo projection too. We git in a little huntin', too."

"The deer all go up higher into the forest this time of year," Rasmusson stated. "You'd have better luck farther up the road."

"Oh, thanks for the tip," replied the man. "We'll try it next year."

"Well, I have to keep moving. Starting to stiffen up." Rasmusson nodded and turned towards the road.

"Hey, you have a great day," the man called after him.

Nostradamus ran out ahead and disappeared into the trees.

Brandon was sitting on the porch, blowing smoke gently out of a rounded mouth and stroking Nostradamus as Rasmusson jogged up. Rasmusson then leaned over and rested his hands on his knees again.

"Notice the altitude?" Brandon asked.

"I'll say," Rasmusson replied, "or last night's pizza."

"It'll take a couple of weeks before you start building up the red blood cells. There's some pills in the cupboard that might help ya."

Rasmusson stood up and stretched his hands high above and behind his head. Then he leaned on the banister, draping both hands over the side.

"What kinda pills?"

"CoQ-10," replied Brandon. "I wouldn't offer you anything outside your comfort range."

The sun was above the mountains; the pines took on a softer look in the warmth. A crow poked at some carrion under one of the trees across the small meadow. Others soon joined it. Nostradamus leapt from the porch, and striding across the drive and meadow, he had soon scattered them amid squawks and a flurry of beating wings. A few high cirrus clouds could be seen suspended in a broad blue sky, which indicated another dry, fair day.

"I ran into your campers up the logging road this morning," Rasmusson finally said.

"Yeah, I guessed that from Nostradamus's reaction when he came back." Brandon dropped his cigarette butt and stepped on it. "A bunch of good ole boys enjoying the glorious outdoors?"

Rasmusson shook his head and looked over at his son. "What are they doing this for? What are they going to gain?"

"Wrong question," Brandon said dryly. "It's their nature; it's what they are."

"It makes me feel so naked and vulnerable." Rasmusson shook his head again.

"Yeah, that's part of it. You'd have never seen them if they really didn't want you to. No doubt those guys left some of their surveillance shit in our cabin."

Rasmusson looked over at Brandon with some amazement but also with fear that he might be right.

"There's a store in the Springs where we can get some equipment, if you like," Brandon added.

"Equipment?"

"Countermeasures," Brandon clarified.

"Oh."

It was then that a large black panel van with dark tinted windows turned up their driveway. It lumbered slowly up the driveway and pulled in front of the porch, and the window slid slowly down. Father and son waited to see what would happen next.

A large man with a scruffy dark beard poked his head out the window. "This the Rasmusson's?"

"Yes," Rasmusson replied. "What do you need?"

"I got a ton a groceries here," the man replied. "Did you order them?"

"Yes, sir," Rasmusson said, smiling. "Brandon, can you show him where to put the stuff? I really need to take a shower; the sweat is starting to crust."

"Sure."

The van was just pulling out of the driveway as Rasmusson emerged from the bedroom, hair still wet.

"What's all this?" he said, surprised. "There must be a mistake."

There were cases of canned goods stacked at the end of the counter, bags of flour, coffee, and sugar, canned and dry dog food, and a stack of toilet paper and other paper goods. The counter was congested with teetering piles of drinks, condiments, vitamins, herbs, and sundry dry goods.

"Don't fret," Brandon said while sorting some things into the pantry. "They called yesterday to confirm your order, and I added a few things."

"A few things? We could keep the 10th Alpine for the winter!"

"Naw, they train down by Telluride nowadays," Brandon responded. "Besides, you're gonna need to be self-sufficient."

Rasmusson's life was so full of surprises and uncontrollable events lately that this seemed like a triviality to him now. Besides, maybe Brandon had a point, although he did not want to think like that.

"That's fine, but you get to find a place to store it all. I'm going to finish up some stuff in my room," Rasmusson said with a mild display of resignation.

"Not a problem."

Rasmusson turned and went back into the room. He sat down at the desk and began reviewing the copious police files that Samuels had sent him. It took Rasmusson a while to organize them in his mind. Samuels had taken many paths in the investigation. It was evidence that Samuels was deeply involved in the case, a confirmation of what Rasmusson had surmised about him. One set of files looked into a radical group, called Anti-SMACH, which espoused the elimination of all sentient machines.

"Too fanatical," Rasmusson murmured to himself. "They would not have the technical expertise to pull it off."

He then noticed that the FBI had interrogated key members of Anti-SMACH well before the sabotage had occurred. The FBI notes indicated a concern that the group had long had a target of disabling a high-profile sentient like Bee. There had been some loose claims by the group supporting the sabotage.

All of a sudden, Rasmusson was struck by one of the documents in the file. The interrogation had reached the level of congressional interest, and a short hearing had taken place about one year before the attack on Bee. The lead interrogator for the hearing was a high-level congressional aide named McLaughlin! He had private conversations with members of Anti-SMACH over a period of weeks, and then the investigation was abruptly dropped by both Congress and the FBI.

Rasmusson leaned back and scratched his head, remembering again the fervor with which McLaughlin had assailed him during the security meeting.

"Search all files for McLaughlin," Rasmusson said. To Rasmusson's surprise, an entire file appeared on the screen:

Henry Mather McLaughlin.

"Ha," Rasmusson blurted out his surprise. The date on the file was two days ago. Samuels had picked up on McLaughlin's vehemence at the security meeting, and it had made him suspicious! Samuels had already compiled a number of interesting facts.

"Gutsy thing for a local cop to take on the Feds." Rasmusson's admiration for Samuels grew. McLaughlin had made the rounds. He was one of those "fix-it" men for a number of government agencies including the Defense Department and the NSA, both agencies with less than scrupulous, but very able, sentients.

Rasmusson's interest was piqued just as there was a knock on his bedroom door and he heard Brandon speak.

"There's another van coming up the drive. You expectin' someone else?"

Rasmusson was torn between his interest in the files and the need to be aware of his environment. Finally, he went out to greet the van while Brandon finished putting away the hoard of food. This time it was a white panel van, but it was yet another scruffy bearded man that poked his head out the window.

"Hi there, we're lookin' for the Rasmusson's ... cleanin' service?"

"This is the place."

"Great. My wife'll come in and look around while I start unloadin' things."

Rasmusson showed the woman around and explained what he wanted. She seemed very efficient and took notes. After listening for a while, she said, "I got my husband and two kids with me. They're hard workers, but this place is gonna take all day. You know our rates. You OK with that?"

"It sounds fine," Rasmusson said, somewhat dismayed that his research would be interrupted. "I'll take my son down to the Springs and be out of your way."

The woman nodded in approval. "Good idea."

Rasmusson found Brandon on the couch eating a bowl of microwaved chili for breakfast. As Rasmusson approached, Brandon looked up at him without expression.

"Do you want to show me that store with all the gadgets you talked about?" Rasmusson asked.

"Sure." Brandon's face brightened. "Right now?"

Rasmusson nodded, and Brandon jumped up, gulping down the last of the chili.

"I'll get a sweatshirt, and we can go."

Rasmusson made some final arrangements with the house cleaners before Brandon emerged in a blue and gold Berkeley sweatshirt. As they were leaving, Brandon held the door so the man could bring in some equipment.

"Be sure and clean up any bugs you find," Brandon told the man as they passed.

"Uh, sure," he replied with a tentative look.

Safe and Secure was a modest-sized shop on the corner of Platte and Wahsatch in Colorado Springs. The sign had obviously been hand-painted by the owner. Weeds grew up between the sidewalk and the building. Inside, it was bursting with pea-sized camcorders, transmitters, software, and various devices. It had a chaotic appearance that was an indication of an owner who had more passion for his enterprise than business expertise. It was clearly one of those shops that existed more because of a loyal, satisfied clientele than good marketing or presentation. It created a comforting ambience for Rasmusson.

"Hey, Ess, you in here somewhere?" Brandon shouted as soon as he walked in the door.

"Hey, Rass, how's yer ass?" The voice came over a PA system. "I'll be out in a lark; get yerself a cup o' dark."

Rasmusson looked quizzically at Brandon.

"I've done a little freelance work for him," Brandon said while walking to the counter and picking up a small, M&M-sized camera and examining it between his thumb and forefinger.

"Totally self-contained, hi-res, lifetime batt, one mile ultra-low RF widge," Escalante said, walking into the room behind the counter. "Just got the first load this week."

He rolled over the counter to slap Brandon on the back. "Where ya been hibering, man? Mizzed ya."

Escalante was one of a new wave of fad speakers, who moved incongruously between idioms and slang in different languages, but he did it with the charm of a mild Mexican accent.

"Yeah, Ess, I've missed you, too."

Escalante smiled broadly, which raised his dark black mustache like a pair of crow wings, revealing the gap in his front teeth. Rasmusson noted the genuine fondness the two had for each other. Escalante rolled back on his feet behind the counter and his demeanor then dropped like a fire curtain as he looked over at Rasmusson, a stranger.

"Ess," Brandon said, picking up his cue, "this is my dad."

"Oh! Hesus-Maria! The famous Doctor Rass!" Escalante's mustache twitched upwards again, this time showing what seemed to

Rasmusson to be about 88 white teeth. "It is such a pleasure!" He leaned over the counter and started pumping Rasmusson's hand with both of his.

"I've been lurkin' at your website all week. Truckload a kaka those SOBs been dumpin' on you. May I get you a cup of coffee, sir?" He dropped Rasmusson's hand and gestured to the coffee machine.

"No, thanks," Rasmusson said, returning the infectious smile. "I would like your professional services, though."

"Ah, mi casa, su casa." Escalante bowed slightly with pleasure and spread his arms out to embrace his store.

"Suits in the bush above the cabin," explained Brandon.

"Capice ya there, Noffaa man." He paused a moment. "Hey, I got a sizzlin' widge in the back that might help." With that, Escalante turned and trotted behind the curtain.

"Noffaa?" Rasmusson asked, looking over at Brandon.

"No fun at all. It's military jargon," Brandon explained with some editing.

"I like your friend," Rasmusson remarked to Brandon, who kept looking at the miniature camera in his hand but smiled a little. Then after a moment, Brandon spoke.

"Ess was an Army Intel security officer for a dozen years, lots of the best training, then he got fed up with all the junk they had him doing. Refused a big offer to re-up. They watch him, too. They watch anyone who leaves the fold; he has to tread lightly."

Escalante returned carrying a black case.

"U-pipe," he said, looking at the father and son. "Universal pied piper. Finds the rats and gets rid of them. You never seen it, OK?"

Rasmusson looked at Brandon, who nodded in assurance. Escalante looked at Rasmusson, who then nodded too.

"Full-spectrum jammer is operational in the back," Escalante started, "really pisses them when I turn it on." With that, he opened the case and revealed what looked like a small communications and command center. He fondled the buttons and sliders and explained its intricacies.

"NetNOS-independent, latest technology," Escalante said, ending the description. He turned his gaze to Rasmusson. "I'll be at your place first thing in the morning, sir. We'll go hunting."

"I don't want to make problems for you," Rasmusson said, leaning forwards and pursing his lips; it was his instinctive way to convey sincerity.

"No, sir! This is what I was born for. Helzbelz, I'm hangin' the 'gone hunting' sign out tomorrow, no matter. I plan to be *je ne sais pas* when those suits come about the jamming violation, anyway."

"Hey, Ess," Brandon interjected, "can we take some of these M&Ms and a receiver?"

"Betcher your bottom, man."

"Here you go," Rasmusson said. He handed Escalante his credit card and placed his thumb on the scanner as if he would be offended with any other option. Escalante took the card and scanned it.

"Ah," he said, looking at his monitor, "no joy. Your plastic has been jacked."

"What?" Rasmusson leaned over the counter to look. "I used it this morning to pay the house cleaners, and it was good then."

"Let me check." Escalante typed on the keyboard. "That infosite from Cornell is vapor, too. Full lemon screw is on. No worry, you're on account."

Rasmusson leaned on his elbows and looked up at Brandon. "They killed my credit already!?" He paused for a moment and then added, "Hmph, at least we have a year's supply of chili. Can't be all that bad."

"It's vegan, you know," Brandon said, smiling.

"Hey, Ess, mucho grass." Brandon raised his fist and bumped knuckles with Escalante.

"Mucho frass, Rass." Escalante flashed his smile.

Rasmusson thanked him as well and turned to leave, and then he turned again.

"Mister Escalante, while I still have some legal tender in my wallet, can I invite you to join us for lunch?"

"Mucho grass, Doctor Rass."

They had a long lunch at a local café. The warm afternoon sun was agreeable enough that they ate outside on the patio. Rasmusson did not have much of an appetite and settled on a Caesar salad with tea, but then he ordered a heavy chocolate cake for dessert. He enjoyed Escalante's company; he especially enjoyed how he drew Brandon out. The chatter was light and carefree. It was a precious moment for Rasmusson, who was able to see his son as a witty, charming adult for the first time. Brandon got up and went to the restroom after the check had finally come.

"He is very happy to see you," Escalante said. "He has been very sad this year, and I have not seen him for a couple of months."

"I think he likes your company; you're his friend," Rasmusson responded.

"No, no, Doctor, I haven't seen him like this for many months; it is you. It is very good you believe in him."

Rasmusson surged at this comment. "I try to support him. I suppose you know he is bipolar, but he won't take his medication..."

"No, Doctor R., that's not it." Escalante leaned over and put his hand on Rasmusson's wrist. "You *need* to believe in him. He has a gift. I have seen it. It is for you, you must believe."

Rasmusson cocked an eyebrow and looked intently at Escalante, who held fast and returned the look. Brandon appeared from the hallway. Escalante squeezed the wrist and made a quick dip of the head to punctuate his point.

Rasmusson paid the check and threw in a large tip. The group left the café.

"Hey bro, don't be a stranger." Escalante wrapped his arms around Brandon and hugged. Brandon stood a little stiffly but smiled.

"See ya tomorrow, Ess."

"Doc," Escalante said, putting out his hand. When Rasmusson put out his, it was grasped firmly again in two hands.

"Thank you for your help, Mister Escalante."

"My honor, sir."

Escalante waved and smiled broadly as he skipped down the street.

As Rasmusson had come to expect, the ride home was shrouded in silence. Only once, as they were riding up the canyon, did Brandon speak.

"Remember when you took me hiking to the top of the Peak? Do you want to go again?"

"I don't think I have the conditioning to make that right now," Rasmusson said. He was taken aback by the suggestion.

"We can drive to the lookout and walk from there."

"It could be cold up there this time of year." Rasmusson was still unsure about this idea.

"It will be a beautiful day; you'll see."

"OK. Let me know when."

As they came up the drive to the cabin, they could see that the cleaner's van had gone.

"Stop at the shed," Brandon said as they got near it. "There's something I want to show you."

Brandon got out and walked over to a dead pine tree across the drive. Reaching among the brittle branches, he pulled out a key.

Nostradamus barked and jumped from the bushes. Brandon took a moment to acknowledge him, grasping his collar and scratching his head.

"I keep the spare key here," he said, returning to his purpose.

He walked over and opened the padlock. They walked into the fusty anteroom of the shed. It had once been a shop. The back part, which was still locked up, had been the garage area. There was a portable generator hooked up in one corner of the shop. On the other side were half a dozen red metal drums. Brandon went over to the drums and pulled out a valve and spigot device from behind one of them. It had a long tube dangling from it.

"This is a hand pump. The barrels are filled with gasoline. All you have to do is open a barrel with the key wrench on the wall there, stick this in, and start pumping. Use the gas cans on the shelf there. The generator holds five gallons. It will run about 20 hours on one tank. It is hooked up to both the house and the shed in case power goes out. The blue plastic barrels next to the desk are water; it's potable. The gray containers over there are kerosene. There are a couple of kerosene lanterns hanging just inside the laundry room. There is a kerosene stove in the tool shed with axes and a chain saw."

Rasmusson peered at the archaic technology. How much easier it would be to have a hydrogen-generating algae pond out back with some fuel cells. Still, he was fascinated by the tutorial but a little bewildered by its purpose.

"What's in the garage?" he asked.

"That's not important to you," Brandon replied. "Grandpa's 30-06 is in the case over the desk. Shells are in the top right drawer with the case's key."

Rasmusson recognized his dad's gun case. He had been taught to use the hunting rifle as a child.

"I am not much into guns anymore," he said.

"Neither am I," replied Brandon, "but I've oiled it, and the sight is aligned. The instructions for the generator are on the plate on the side. I think that's it."

"You're not going to show me your horticulture enterprise in the garage?" Rasmusson asked.

"If you like," Brandon replied, "but I don't think it much matters."

"It could be an excuse for them to hassle us. I assume it is an unlicensed crop?"

Brandon smiled. "Licensing means regular inspections and cuts into profits. Anyway, they don't need to have that excuse. They can generate any reason they need, if they want to hassle you. You know that."

"For the record, I disapprove."

"Not a problem," Brandon responded. "The last harvest has gone to market, and I don't plan any new crops right now. If you happen to need it, there are some operational reserves in a tin in the bottom drawer of the desk. Questions?"

"Maybe later. I'm tired right now. I am going to go take a nap."

They locked up the shed and drove up to the house. It looked entirely redone. The windows were clean, the curtains obviously vacuumed. The walls were several shades lighter, and the wooden floors brightly waxed. It had that newly laundered smell. Rasmusson looked around with pleasure.

"They did a good job," he said.

"Yeah, a very industrious lot," Brandon said. "One more thing. I'm going to move in with Jennie. I've talked to her, and it's all set. I am going to pack up tomorrow after Ess comes."

Rasmusson looked at him and gave a half smile. He was too tired to offer any other response. He walked into the bedroom, kicked his shoes off, and collapsed onto the bed, pulling the pillow over his eyes and quickly falling into sleep.

About midnight, Rasmusson heard the dog barking. It was a much different bark than usual. It contained a fierce quality, the instinctive reaction of an animal protecting its pack. Then it quit suddenly. Rasmusson rolled over and fell back asleep. Later that night, he rolled to the side and sat up on the edge of the bed. The room seemed to wobble, so much so that he had to steady himself by grabbing the headboard. He felt a pit swelling in his stomach and then the nausea. Quickly, he stood up and stumbled across the room into the bathroom, where, kneeling at the toilet, he threw up what little was left of the Caesar salad. When he was finished, he sat down on the floor, holding onto the toilet bowl. He rested before pulling himself up to the sink and washing off.

He then went back to the bed and sat on the side with his hands supported on his knees. He thought he heard some sounds, a little bit like sobs, coming from the living room. He got up and walked out.

"Sorry you're awake." The voice came from a dark, indistinguishable figure sitting in the corner.

"I didn't mean to wake you," replied Rasmusson. "The salad must have had some bad dressing or something. It certainly was not as good the second time."

Brandon said nothing, and Rasmusson sensed something was wrong, so he just stood there, waiting for a response.

After a weighty silence, it came.

"Nostradamus is dead."

"What?" replied Rasmusson, taken aback. He had become fond of the dog and also recognized how strongly his son had felt towards him.

"What happened?" he asked after a pause.

"His throat was torn out," Brandon replied quietly. "I heard some noise and called him. He didn't come, so I went looking. I found him down by the shed in a pool of blood. He was barely breathing, and even that was through the gashes in his neck. When I picked him up, his head rolled back and blood poured out of his nose. The neck was broken. So I got the gun and ended it for him. He is buried on the right side of the driveway. Sorry if I woke you. I tried to muffle the sound with a blanket.

Rasmusson was amazed that he had missed a gunshot, muffled or not.

"Was it a mountain cat? What did you hear?"

"Nothing," Brandon replied. "There were no slash marks like a cat would leave. This thing went right for the neck and throat, more like a trained dog, maybe a wolf. It must have been a strong bugger to break the neck like that."

"Have you called animal control?"

"If you don't mind, I'd rather just be alone right now," Brandon replied tersely, turning to go to his room, "and you look like you need to go back to bed."

Rasmusson stood motionless in the middle of the room as Brandon got up and walked towards the bedroom.

"I am sorry, Brandon. I liked Nostradamus."

Brandon glanced back and nodded in acknowledgment before climbing the stairs. Rasmusson went back to bed, but he did not fall asleep so quickly this time. He kept seeing the image of a limp, dying dog in his son's arms like a thick black throw rug on a clothesline. Then in the back of his mind he would hear a bitterly ironic voice repeating, "a boy and his dog."

The sun came late the next morning. It was after 10:00 when Rasmusson threw back the cover and climbed out of his bed. A little while later, he emerged from the bedroom. Escalante was sitting at the table with assorted equipment, including an attaché case and a number of small, peanut-sized devices. Brandon sat motionless on the couch

"Ah, Professor, good morning!" Escalante greeted Rasmusson. "Fine ento collection you have here."

"Ess has been busy this morning," Brandon said.

Rasmusson walked over to the table to get a closer view. Escalante held up a one-inch long tubular peg with wires hanging out.

"Found this guy drilled into the bottom of your chair leg and plugged with wood putty. He's the master switch. Detects whether a bug search is on and turns off the colleagues. He sends out a periodic pulse. If they don't get the code, they turn off. Makes 'em damn hard to find. Some of them come on periodically anyway. It is all pretty sophisticated."

"So how did you do it?" Rasmusson asked

Escalante smiled and laid his hand proudly on his case. "Picked up on mister master here first, encoded the pulse, and continued to send it out. Then we went for the little bombers..."

"Ess knows his stuff," Brandon added. "He probably got them all."

"No man, not yet," Escalante added with emphasis.

"Point is," Brandon said, continuing to look at his father, "there are some people with deep resources who want to smell every time you fart."

Rasmusson was struck with the similarity of this with something that Clavell had said at the Ithaca College meeting.

"What am I supposed to do about it?" Rasmusson felt a little annoyed.

"You're the smart one," Brandon said and got up to get some coffee.

"Hey, Doc," Escalante said. "I need to look some more outside. Do you want to join me?"

"I'll get cleaned up and have some food, then I'll find you outside."

Rasmusson poured himself a bowl of Wheat Hearts and milk and put them in the microwave. He looked over at Brandon, who was curled up in a lotus position on the couch, gazing out the window towards the shed. Rasmusson ate his cereal without comment. He saw the mound of freshly turned earth that Brandon was looking towards.

When Rasmusson finally walked out on the veranda, he found Escalante on an extension ladder leaning against the eaves. Only his legs were visible.

"The gutter needs cleaning, while you're up there," Rasmusson said.

"Yes, sir, I am finding a lot of junk up here," Escalante replied and then dropped his arm down to where it was visible. He held one of the M&M-sized cameras between his thumb and forefinger.

"They have these all around. Found 14 of 'em so far. You're on candid cam. Top-quality shit, man."

Rasmusson felt that now familiar feeling of being naked and violated sweep over him.

"I think this is the last one," Escalante continued as he descended from the ladder. "I'll need to come back in a couple of days and do another sweep."

"Mr. Esc—" Rasmusson paused. "Ess, how can I repay you?"

Escalante walked up the steps and put his hand on Rasmusson's shoulder, fixing his gaze. He spoke with such uncharacteristic deliberateness that it caught Rasmusson's attention.

"Here's the deal, Professor," he said. "Conspiracies are not as well-organized and efficient as they want you to believe. It's like any other project with a bunch of people; they have screw-ups, miscommunications, cross purposes and all that crap. But, one thing they do well is make the victim think they have some unseen power, an' ya can't do anything against them. All these electronic turds they left around do that. That is sometimes more important than the surveillance itself."

"They want to intimidate me?"

"No doubt," Escalante answered. "Some of these bugs were meant to be found. Not all, though. Some of it's damned modern stuff. Lotta capital investment here."

Escalante noticed that his speech had left Rasmusson looking crestfallen. "Show 'em your stuff, Doc. That's the best thing you can do."

Rasmusson looked up. "I'm feeling a little fatigued. That's all. I'm OK."

"Ya got some good help, ya know. Professor, remember what I told you at the café about Brandon's gift?"

"Yes, I appreciated ..."

"Professor Rasmusson," Escalante interrupted him, "this is also part of the deal—you must not take Brandon's moods personal. He loves you and respects you. I knew this a year ago, before I met you. That is part of the reason he treats you the way he does. Be patient with him; he has a heavy burden to carry."

Rasmusson stared back into Escalante's dark eyes and saw a new depth to them. This eccentric Hispanic, who was such a good friend to his son, now amazed him with his wisdom and insight.

"OK," he replied, nodding his head, "it's a deal."

"Ah, superissimo!" Escalante smiled broadly and pumped Rasmusson's hand. "Hey, Brandon," he shouted through the open door. "Adieus on youse. I shall return in a couple."

Escalante skipped down the porch, threw his gear in the back of his pickup, and then hopped into the cab. Brandon came out the door and made a simple gesture, but his affection for his friend was revealed in his expression as he watched the truck rumble down the driveway. Rasmusson stared at this.

"What?" Brandon had noticed his father gazing at him.

"Oh, nothing." Rasmusson turned and looked towards the sky. "I like your friend."

"Of course," Brandon said as he walked back through the door.

Rasmusson stood against the railing on the porch and enjoyed the view. The Peak, which was a good seventy miles away, was as sharp and detailed in the clear dry air as the sapling pine next to the shed. Small whitish dots on the horizon indicated clouds hundreds of miles away.

"The sky is so much larger in Colorado than New York," he thought to himself, and then another unwelcome thought crept upon him.

"The satellites," he muttered under his breath as he turned and went in the house to escape the light of day.

The house felt more comfortable to him now. It was similar to when the house cleaners had finished their cleaning job. Brandon was stretched out on the couch, a pose that was so common when he was growing up. Rasmusson stared again.

"You know," Brandon said, looking up, "you're bein' just a *little* weird today."

"Yeah," Rasmusson replied, "I know."

Rasmusson grabbed a can of cranberry juice, taking it into the bedroom where he sat down at the terminal and flicked it on. What he saw startled him. Instead of one of the usual screen starters, there appeared a strange configuration of boxes on his screen with white, gray, and black checkerboards.

Rasmusson looked at it with irritation. The investigation was on his mind, and he was anxious to do some research on it. He rebooted the machine, and the image came up again. Out of frustration, he rebooted again. This time, he looked at the pattern a while, and then his native curiosity overtook his annoyance. He started mumbling to himself as his mind impulsively began unraveling the puzzle.

"Some kind of encrypted code," he thought. "Hmph, every block is three deep with three shades—white, gray, or black. That's 27 possibilities, enough for the 26 letters—this shouldn't be too hard to decipher. Any single-column block has to be 'a' or 'I.' Not many choices for two columns. They have to correspond with 'to,' 'an,' 'be' . . . 'The' has a high Zipf probability for three columns. . . "

Within five minutes, Rasmusson had the following scribbles on a pad:

"and, I, am, *ith, you, . . . to, the *nd, of, the, a*e"

"Ha!" he exclaimed suddenly. "I know this!"

He got up and walked over to a bookcase. Stooping over, he found an old, heavily worn Bible. He thumbed energetically through the concordance, while the dust flew up, and then he turned to the last verse of Matthew.

"And behold, I am with you always, even unto the ends of the age."

"Ha ha!" he again exclaimed with great pleasure. "It fits! All those early years in Sunday school and Bible study weren't such a waste."

He was about to punch through the standard set of rescue keys when something about the geometry of the screen engaged him. He hit the Print Screen button. As soon as he had done so, the mystery screen disappeared, and his normal startup screen reappeared. The printer spewed out the image on paper, which Rasmusson then taped to the side of his desk.

"That was odd," Rasmusson thought to himself, and then he went back to his task.

He opened his email. There was a conspicuous lack of email, even from Ellie.

He then opened the scanned police files from Samuels and began poring over them. Samuels had left a yellow sticky note on the section of the McLaughlin committee that said simply, "On hold, see Chief."

What did that mean? Did Samuels get directions from superiors telling him to quit that direction of the investigation? Why did Samuels bother to scan the yellow note with the other files that he sent?

It wasn't long until Rasmusson's eyes grew heavy and his head started to dip. It was becoming a habit to kick off his shoes and slide from his chair into the adjacent bed. He slept the rest of the afternoon. The last thing he remembered before he woke were the images of white, gray, and black dots flowing through his veins.

He never used to nap, and these naps were not restful. He woke up feeling drugged and mildly nauseated. He went to the bathroom, washed his face, and held a wet rag on the back of his neck. The countenance gazing blankly at him in the mirror surprised him. It was an old man's face with deep, hard furrows, pale skin, and blue bags that wilted under his eyes. The eyelids were mottled with tiny red spider veins.

He felt hunger, though no particular food appealed to him. He lurched out unsteadily into the living room. It was dark. It was quiet. Some time in the afternoon, Brandon had packed up his things and left without a word. It now felt darker, quieter, and lonelier than Rasmusson had known it. He wandered over towards the refrigerator to get a cola in the hope that it would wake him up. On the way, he stubbed his toe on the leg of a chair. He stood for a moment, absorbing the pain, and then continued plodding towards his goal. It seemed to him that the pain in his toe grew. He could not handle pain as well anymore.

He was not surprised that Brandon had left, so he felt no right to be disappointed. He reached and grabbed a two-liter bottle with some dark liquid. The sugar would feel good. As he lifted the bottle to his mouth, he noticed a note scribbled on a napkin under a magnet on the refrigerator door:

Gone to Jennie's
Will check in occasionally

B.

"He left me a note." Rasmusson's spirit was raised; the room seemed noticeably brighter.

THE SUMMIT

. . . and crown thy good with brotherhood
from sea to shining sea!

—Katherine Lee Bates,
"America the Beautiful"

The Indian summer dragged well into October. The nights were crisp and cool, but the days remained pleasant. Bright yellow swatches of aspen appeared amid the dark green pine and quilted the mountainsides. Winter waited just offstage.

Rasmusson was aggravated. He would sleep late and take long afternoon naps. In between, he would spend long periods sitting on the veranda, staring vacantly at the forest. At first he had thought it was the altitude and stress, but it was deeper than that. He thought of going to a doctor but did not know anyone he could trust. He could no longer trust strangers to poke him with needles and examine his body fluids.

The long sleeps did not relieve the fatigue. He was constantly tired. The only thing that seemed to move him was hunger and some critical body functions. He had not been jogging since the day he ran into the "campers." He thought of the file from Samuels often, almost longingly, but could not raise the energy to sit in front of it for more than a few minutes. His vaunted ability to concentrate had abandoned him. Sometimes he would search his email, but there seemed to be nothing of interest. Even Ellie did not write anymore.

It was foresight on Brandon's part to show Rasmusson the generator and how it worked. The night after Brandon left, all power to the cabin went dead. He now ran the generator for two hours in the morning and in the evening. It was enough to handle his needs and keep the refrigerator and freezer contents properly cold, if he did not open them in between. It was not a problem for Rasmusson, who was quite happy to sleep in between.

Several nights after Brandon left, Rasmusson sat in the dark on the porch wrapped in an afghan waiting for the generator to run long enough to recharge the refrigerator. He thought he saw a figure of a man scurry between the pines down by the shed. Suddenly, the shed light, which was connected to a motion detector and which Rasmusson had deliberately activated, flashed on. He thought he saw a couple of figures disappear behind tree trunks. Rasmusson was unnerved and went inside, turning off the lights and shutting the curtains. This happened again the next night, again while the generator was running.

The third night when the phantom visitors appeared, he heard banging noises. Someone was breaking into the shed. This time, instead of hiding in the cabin, he sighed deeply, got up slowly, and walked nonchalantly down the drive to the shed. After all, what would they do to him? If they wanted to do something physical, it would be easy at any time. When he got to the shed, he found that he had been right. The padlock lay on the ground, severed by a bolt cutter, but the culprits were not in sight. He gazed around slowly at dark voids in the trees. He methodically opened the door on the shed, went in, and turned on the shed light. He looked around, but nobody was inside. He must have interrupted them before they could finish their job. A minute later, he emerged with the 30-06 cradled in his arms. His father's old gun gave him a feeling of comfort and strength. He had opposed guns most of his adult life, but now he had a feeling of consolation. The gun would not really protect him, but it made him a little more menacing, like the thorn of a rose about to be plucked. The shed light flooded the woods with that blue light that a mercury-vapor lamp creates. Long shadows fell away from the light. He raised the gun to his shoulder and fired a round into one of the nearby trees. The snap of the gun sounded throughout the forest and careened off the nearby mountains. He then pointed the rifle at a hollow log and fired off a succession of rounds. The log acted as a sounding chamber, cracking loudly every time it was hit. There were footfalls among the trees. Rasmusson fired a couple of more rounds into the log. After waiting a moment, he turned around, flipped off the generator, and locked up the shed with another lock from inside the shed. He took the rifle and a box of shells back to the cabin with him. There were no more incidents at the shed after this.

Brandon had left him a good supply of chili, but eating it, Rasmusson found, would cause a rash and a relentless, merciless itching. Anything with peppers of any type in it now did. It was one of the few things that would wake him up at night. During these episodes, half conscious he would claw at his back and legs until, one morning when he awoke, he had left bloodstains on the sheets. After that, he adopted the habit of taking large doses of antihistamines and cortisones before bed. He suspected he had become sensitive to milk, too. He had heavy congestion and choked up heavy phlegm in uncontrollable coughing spells. He often got debilitating cramps in his legs and back muscles whenever he consumed any dairy products. He consumed a lot of aspirin and related pills.

Once, he dipped into Brandon's slush funds, made it to town, and bought a few supplementary items, herbs and medicines he thought might work on his many new ailments. He took all the esoteric supplements with odd-sounding names: ginseng, dong quai, wild yam, Peruvian cat's claw, and CoQ-10 with a large vitamin pill that the herbalist had said would solve his fatigue issues. He threw it all up.

Escalante did not come back as he had promised. Neither had he heard from Brandon for three weeks. In fact, he had not heard from any family, friends, or colleagues. It was strangely quiet. In his few alert moments, he could imagine this as part of a conspiracy. He sensed the depth of the powers that had arrayed against him for weeks now, but if this was persecution, it was not what he expected. Persecution, he thought, was shrill and raucous. It was bricks through windows, cross burnings, threatening phone messages, mobs, and public denunciations. What he experienced was isolating, silent, bewildering, and lonely. It was more like he was being actively forgotten, pressed into languid oblivion among the immovable peaks of Colorado. It was not how he expected it to be.

His depression seemed at its worst in the mornings. Some mornings, he would stay curled in a ball in bed, hoping that he could fall back to sleep; but that slumber which came upon him so easily these days would fail him in the morning, so he lay awake glaring at the ceiling. Other mornings, he would let the generator run longer so he could watch a movie. He would watch comedies in the hope of feeling humor, or tragedies in the hope of feeling sadness. Strange, he would think to himself, how he had always thought of depression as sadness, but in reality it was the lack of emotion, the total lack of feelings and urges. He would be happy if he could feel sad. He would be happy to feel like he was anything more than a fleshy vessel of fluids and bones, which kept functioning in a state called life, even if there was no feeling in it. He was not sure he would even know any terror if that thing called life chose to quit on him. Such were his thoughts, when he had any.

One morning, Rasmusson was awakened early by a loud knock on the front door. He jumped out of bed so quickly that he had to prop himself against the desk when he nearly passed out. He pulled on his jeans and threw on a sweatshirt that was lying on the floor. He got to the door in time to see the panel express truck driving away.

An express package leaned against the screen door. It was somewhat longer than standard letter size. Rasmusson stabilized himself against the door frame and stooped over to pick it up. There was a skip of excitement at receiving a package. He came up slowly, carefully.

Inside, he ripped off the cord to release the contents, which he let slide onto the table. There were several legal-sized documents in plastic covers and a letter folded over neatly by thirds. He picked up the letter and unfolded it. Rasmusson recognized Rose's handwriting and the scent of her perfumed stationary. He sat down to read it.

> *Dear Andy,*
>
> *First of all, I have missed you. I hope you truly believe this. I hope you and Brandon are getting along and having a good time and that you are well. Give Brandon my love.*
>
> *By now you have probably already seen the legal papers that came with this packet...*

Rasmusson stopped reading and looked up at the plastic-covered documents. He squinted and focused hard to see what the papers were. Divorce papers. He paused for a moment to let his emotions catch up to this information, but they didn't arrive. He mused over his lack of feeling; he was not sure what emotion he would have otherwise and thought that maybe this was a blessing of sorts. He went back to reading.

> *...I know that my good wishes might seem insincere at this point, but you must believe that I still see you as a good friend. I respect you and have good memories of the many years of companionship we had together. I hope that memory is the same for you. Overall you have done your duty as a husband and companion, and I am very grateful. My*

attorney has said I should not even tell you these things, so take this as a token of my sincerity.

Having said that, you know that people and attitudes change. The recent events and your unwillingness to consider alternatives have upset me terribly. I think you know that, but honestly that is not the reason for my decision to seek divorce, it has only focussed what I think we both know that we really don't have anything in common any more. Our marriage hasn't had meaning or communication or purpose for years. I think you will agree that what my attorney has proposed in the documents is fair. I hope we can remain on friendly terms. I would feel bad if I lost your friendship. Dear Andy, please see it as an opportunity. Such changes often result in grand new starts in life!

Best regards,

Rose

There was a postscript.

P.S:

Please take the time you need to consider this, and perhaps talk to an attorney. If you need to call and talk that is fine. I hope you are getting lots of good writing done on your book and enjoying the mountains.

"A handwritten letter; a nice, human touch," Rasmusson thought. "She always had a sense of these things."

Rasmusson smiled a little. The book, he had not even considered it. Rose was right about one thing: they really did not communicate anymore. He could easily imagine that the implications of his situation and all of the news surrounding it would have passed right by her. She genuinely did not have interest in such things. At times, he had even hoped for some scraps of empathy from her. Rasmusson felt no real hurt. He thought of how natural a consequence this was for him and Rose and how well it seemed to fit into his life at this point. In fact, if Rose were not asking for a divorce right now, then somehow, somewhere, the master scriptwriter would have missed something important in the tragicomedy of his life. He dropped the letter on top of the documents and walked outside. He sat there for a long time.

Later that afternoon, he fixed a bowl of oatmeal by pouring apple juice on it. It was simply prepared, and in its simplicity it seemed to be free of any problems for him. It was simple food with no pretensions. He had even begun to like it. He took the bowl and sat down at the monitor in his bedroom, which ran on its own battery between generator sessions. He scanned the news. There was one item on the so-called Christian virus showing up everywhere. It had managed to evade even the best virus scanners. Fortunately, it was innocuous, simply leaving a biblical quote in a pictograph code and then disappearing. Experts, the report said, were impressed by its cleverness, although security specialists were greatly chagrinned by its effectiveness at penetrating even the most protected systems. It would not be long until the culprit was found, the report said. There were others who had taken an interest in its expressive, two-dimensional nature as a writing form. This reminded Rasmusson of something vaguely familiar, but in his present stupor, he could not quite remember what it was. Rasmusson looked down at the Bible, which still lay on the floor by his bed where he had set it a couple of weeks earlier after he had deciphered the code with it. He sneezed suddenly and violently and had to reach for a tissue. His nose started bleeding. Not all that uncommon in the dry, thin air at this altitude.

Rasmusson awoke in the late afternoon. It was a rare moment where his mind felt lucid. He even felt some remorse arise about his marriage. Rose had been a stunning beauty when they first met. A friend of the girl he had been dating. She was totally at ease in socials, a charming conversationalist. To Rasmusson, she came from a world that he had only read about or seen in movies. For Rose, this intelligent, innocent scientist was fascinating. It made a perfect setting for a movie. Both were mature enough to realize that the mutual fascination would wear down, as all romances do. They settled into their routines. Rasmusson felt pity for Rose; she was a community activist, always involved with charities and activities. He was a zealous scientist, always trying to read some research articles on Christmas Eve before guests came, sometimes even after they arrived, if he found the guests uninteresting. He remembered working on a grant proposal on his laptop during one PTA meeting. What else could he do? That's who he was, but he had more understanding for her. She always had a disgusted look and some sarcasm, but it was no worse than that.

Why had he spent the last year being careful around Ellie? She was so easy to talk to. Her mind's gears meshed with his. Rasmusson admired her talents and determination, but it was the lockstep of ideas and imagination that attracted him to her. Thoughts flowed between them without resistance, a resistance that he had grown so accustomed to with Rose. With Ellie, it was like breathing fresh air after being in a stuffy, smoke-filled room. It occurred to him that he had been looking for her his entire life. He just did not know it until he found her. Rose may have initiated divorce, but it is never one-sided. She may really have gotten it right this time. Rasmusson felt something for Rose. His mind could not put a word to it, but it was a feeling of well-wishing.

Rasmusson's musing was interrupted by sounds coming from the kitchen. He quickly rubbed the sleep from his eyes and reached for the rifle that leaned against the bedstead. He walked to the door and cracked it open slightly. His pulse surged. Through the crack he spied a shock of red hair bobbing behind the counter.

"Brandon, is that you?"

Brandon stood upright and turned around with a bottle of cola in his hand. Rasmusson leaned against the door frame with the rifle in one arm, the barrel pointing to the floor.

"Doin' some hunting?" Brandon asked, looking at the gun. "You look like hell."

Rasmusson leaned the gun next to the nearby couch. He looked at Brandon. His sweatshirt was limp and dingy like it had been slept in for some time. His hair was knotted in greasy mats. His socks hung limply around his ankles

"Hmm," Rasmusson said, smiling, "you're welcome to use my shower, if you still know how to operate one. I'll start the generator, and we can have a hot meal. Run the washing machine too."

"I'll do the shower later," Brandon replied. "Right now, I'd like to get some food."

"There're cold cuts in the back of the second shelf," Rasmusson said, gesturing with his hand. "Make a sandwich."

Rasmusson had thought Brandon looked leaner than usual. Rasmusson then stretched out on the couch and rested until Brandon had eaten something before he spoke again. This was acceptable to Brandon, who quickly put together a meal.

"How's Jennie?" Rasmusson asked, finally breaking the quiet.

"No idea," Brandon replied and then gulped down some of the cola. "She wouldn't let me in. Just kept sayin' I should go away. So I did."

"Where have you been?" Rasmusson asked.

"Explorin' the backwoods, sleepin' in the Burb, doin' a little writing."

"Escalante never came back," Rasmusson said again after a minute of silence.

"Yeah, his place is boarded up. Windows painted white."

"Geez," Rasmusson exclaimed, "I hope he didn't get in trouble because of me!"

"Naw, not him," replied Brandon. "He knew well enough what he was doin'. He'll be OK. Probably with family down in New Mexico."

Brandon's last statement evoked thoughts of Escalante with some amazement. Had Brandon really known that this would happen? That he would have to go away? It was all so obvious and casual to Brandon.

Rasmusson was now at a loss as to what he should say next to keep the conversation viable. After another couple of minutes of silence, he finally spoke.

"Your mom wrote. She wants a divorce."

This news actually seemed to pierce Brandon's defenses slightly. He stopped eating and looked at his father. He tightened his lips almost imperceptibly and nodded affirmatively. It was as if he had resigned himself to the fact earlier, as if Rasmusson had not paid the electric bill and someone told him the power had been turned off. Brandon finished his sandwich, carefully cleaned up the crumbs with his napkin, and threw them into the trash.

"OK," he said, starting for the bedroom, "I guess I will take that shower."

He stopped at the door and turned towards Rasmusson, still on the couch.

"Dad," he said, "do you wanna make that trip to the Peak tomorrow?"

Rasmusson looked up at him through half-closed eyes. He did not want to deny this one gesture of his son, but the thought wearied him. He lay on the couch with his mouth open but with no words.

"I'll drive," Brandon continued insightfully. "We will take the road to the top. You can rest on the way."

Rasmusson hesitated still. The thought of the cold and wind above 14,000 feet enervated him. All he really wanted was a warm bed these days.

"Remember when we used to throw rocks off Devil's Drop and count how long it took til we heard 'em hit," Brandon said nostalgically. It was so completely out of character that it took Rasmusson by surprise.

"OK, you're on," he agreed.

"We can wait until late morning, when it is warmer. S'posed to be good weather tomorrow," Brandon said, dropping the last remarks as he disappeared through he door.

"OK." Rasmusson forced out the syllables with whatever enthusiasm he could muster. "OK," he repeated again quietly to himself.

The rest of the day, when Rasmusson wasn't napping, he spent talking to Brandon, who seemed very interested in events and in his father's poor health.

"I know a naturopath in the Springs you can trust," he said. "I will leave his contact info on the fridge."

"Why, where are you goin'?"

Brandon ignored the question.

They talked about home and growing up, and they also talked about Rose. Brandon, who seemed indifferent about the world, showed surprising sympathy to his father on the subject of his failed marriage.

Finally, as dusk fell, Brandon got up and said, "I will go turn on the generator. You're tired; get some rest."

With that, Brandon picked up the rifle and checked the chamber before going out the door.

It was after 10:00 the next morning when Rasmusson awoke. He was still on the couch, but he had been covered with an afghan and his head was on a pillow. Brandon was at the kitchen counter eating a stack of pancakes with maple syrup. Rasmusson pulled himself up and made it to the bathroom where he took a long, hot shower, grateful that Brandon had turned on the generator early enough to make some hot water. Later, he emerged from the bedroom with a turtleneck shirt, sweater, and ski parka folded over his arm.

"Just like Saturdays when you were a kid and your Mom would fix pancakes, huh?"

"Yeah," Brandon replied dryly, "I guess so. There's more batter in the bowl, if you want some."

After breakfast, Rasmusson filled some water bottles and gathered up his clothing. Brandon was still wearing the same sweatshirt as always.

"Don't you want a coat?" Rasmusson said, reverting to his parental instincts. "It will be below freezing up there this time of year."

"Nope. I won't need it."

Little was said between father and son until they arrived at the toll gate for the Pikes Peak Highway, a euphemism for a dirt road that rose over 6000 feet with more than a hundred hairpin turns before it arrived at the summit. They started up the road winding around the lake and through the pines. At 10,000 feet, Rasmusson started to feel a headache and winced. Brandon, who had been stealing glances at his father while pretending to look at the view, noticed.

"There's some aspirin in the glove box," he said, gesturing with his hand. "It'll help."

"Thanks." Rasmusson opened the glove box, grabbed the bottle, poured out three pills, and swallowed them without water.

"I'm sorry," Brandon went on. "It was really a stupid idea for me to bring you."

"No, Brandon," Rasmusson replied. "It's a good idea. It is really the first sociable thing I have done since . . . well, since you and Escalante were at the place. It is nice to feel like I have a bit of a life. A headache is small potatoes."

"Stupid!" Brandon seemed to ignore Rasmusson's reply altogether. "But I wanted to tell you some things, and I just couldn't bring myself to do it earlier."

"Sure." Rasmusson recognized it was time to listen.

"Life's a bitch," Brandon said, starting to open up. "I don't need to tell you that—hell, you look shittier than me. You and Mom were good parents; I really want you to know that I feel that way. You can tell Mom I said that too, even if it's at your deposition."

Brandon smiled.

"Brandon," Rasmusson said, "what is between your mom and me has nothing to do with you..."

"I agree." Brandon broke up the rest of the thought. "I've managed to screw up my life on my own. Please don't bring up the meds right now; I know that's what you're thinkin'."

Brandon was right. Rasmusson swallowed his words.

"It's deeper; I'm fundamentally broken. It's like I am an alien on this planet. Even when I was on the meds, I never felt at home here. I never wanted to hurt you or Mom, or even be the screw-up I am. Please believe that."

Rasmusson felt emotion swelling and wanted to embrace his son.

"Brandon . . . "

"Dad," Brandon interrupted and paused for a moment, "I never told ya, but I am proud to be your son. I am proud for what you stand for. Don't let them bastards beat you down! They're not through tryin', ya know. They are clever bastards. OK? Don't let 'em win! Don't give into them! Ya gotta promise."

Rasmusson noticed that Brandon was repeating himself, so incongruous to his concise style. He looked over at his son. His demeanor was not the wild-eyed look he expected. Brandon seemed calm and resolute as if he were looking into the distance and seeing something.

"OK, Brandon, it's a promise."

"Promise me," Brandon repeated, "you'll never let them sons o' bitches win!"

"You got it, son."

"I love ya, dad," Brandon whispered quietly under his breath.

"Did you say something?"

"Tell me about the Peak," Brandon said, "just like when I was a kid."

"I thought you would be tired of all my old stories."

"Humor me," Brandon replied.

Rasmusson complied. "Well, OK. On a clear day from the top, you can see the flats of Kansas to the east and the smog of Denver to the north. That dark green swatch down there is the Black Forest. Grandpa's old ranch was just northeast of that. To the south, you have the Sangre de Cristo Mountains. That's Spanish for the blood of Christ. The tall one is on the New Mexico border."

Rasmusson kept talking until the altitude sickness gave him a little nausea. Then he took a drink of water and leaned his head back, breathing deeply through his nostrils.

"Why do they call mountains 'Blood of Christ'?" Brandon asked. "I don't see the connection."

"Not totally sure," Rasmusson responded, "but I can imagine a metaphor between the way in which those mountains reach to heaven and the blood of Christ as bringing one closer to heaven."

"Do you believe there are situations where blood can atone?" Brandon asked. "I mean help make something right?"

Rasmusson looked over at Brandon with some amazement. In spite of all his curiosity, they had never had religious discussions.

He then answered, "I'm afraid I haven't considered such questions enough to give you a good response."

There was a pause in the conversation. The large van kicked up some dirt as it made a turn. Several rocks continued bouncing down the mountainside.

"But I guess atonement and sacrifice make for one of the most dramatic story lines possible," Rasmusson added as a postscript.

The rest of the trip went by in silence.

The top of the Peak was strewn with fractured granite boulders. A rambling gift shop and the landing for the cog rail were situated on the north side of a flattened area. As Rasmusson got out of the van, a brisk wind caught the hood of his parka and slapped it up the back of his head. Rasmusson pulled on a gator around his neck.

"Aren't you cold?" he yelled across the hood of the truck to Brandon.

"No."

Rasmusson then walked over to the eastern platform where a large American flag whipped in the incessant wind. Strains of "America the Beautiful" played loudly on a sound system over the sound of the wind. A bronze plaque commemorated the song and its lyricist, who had been inspired to write the words while on a visit to Pikes Peak.

"Attention," a voice broke into the music. "The cog rail will be leaving in five minutes."

The scratchy music resumed. Rasmusson felt a hand on his shoulder and turned around. Brandon leaned over and shouted toward Rasmusson.

"Why don't you get some of those high altitude donuts for us from the shop? I'll be on the west edge by Devil's Drop. You know, where we used to throw rocks off without bein' seen."

"Good idea," Rasmusson said, smiling at the reference to rocks. He yelled back. "A little sugar might help the headache."

Brandon smiled.

The shop was crowded. The heat and the shelter from the wind felt good. Rasmusson's cheeks began to warm up. The trip had taken his final dram of energy, but he thought how he would curl up on the back seat of the Suburban with the flannel blanket and sleep going back. It had all been worth it.

Between the chronic fatigue and altitude sickness, Rasmusson had a frightening pallor. His eyes were sunken, and he stood unsteadily while propping one hand on the counter.

The clerk selling donuts looked at him sympathetically. He had often seen people with altitude sickness, but Rasmusson looked much worse.

"You all right, sir? I can get some oxygen."

"Yes," Rasmusson said, looking at the clerk and managing a slight smile, "I feel good." And then he added, "In spite of all appearances."

The clerk went back to counting out the order of donuts. Rasmusson thought to take some back to the cabin, even if they would shrink up like a wad of cellophane in the lower altitude. They still tasted good. Later, he and Brandon could make hot apple cider, sit by a fire, eat the donuts, and talk some more—maybe more about religion. It sounded like a good plan to Rasmusson.

At that moment, a man came running into the shop, slamming the door open and gesturing wildly toward the west side of the shop. His eyes were violent with stimulation and his face was flushed. Everyone in the shop turned to look at him.

"Call 911! Get some help!" he yelled. "Some kid just jumped off Devil's Drop!"

Rasmusson wheeled around to see. The middle-aged man had long gray hair tied in a ponytail and wore a black leather jacket.

"Redheaded kid," the man said, still gesturing. "He just walked to the edge, looked back once, and walked off."

The last thing Rasmusson remembered seeing was a tattoo on the man's right hand with an anchor and bold yellow letters on a red banner that said "semper fi." Then everything went dim. He lost consciousness in a world of murky shadows.

CONVERSATION

The snow fell lightly, only a few fluffy inches, but it covered lawns, roads, leaves, and rooftops in a silent veil under soft white skies. Rasmusson's hospital room window looked out upon the wintry view towards the Peak, but the mountain was obscured in clouds.

Rasmusson had not been conscious for over a week. He had been vaguely aware of a heavy darkness that enveloped him, punctuated occasionally with indistinct images. He had seen doctors and nurses, felt pointed things pricking and piercing his flesh. Once, he thought he imagined seeing Rose, lifting a dark veil, leaning over and kissing him on the forehead while a large man in a dark suit stood behind her. Another time, he saw a trinity of imposing men standing at the foot of the bed. He was sometimes aware of a heavy weight, brooding in his chest, but then he would drift mercifully back into the darkness, darkness without rest or peace, but also without sentience.

He finally awoke to see a nurse standing next to the IV line with a syringe held needle up. He blinked a couple of times as she came into focus. The first thing he noticed was that she wore scrubs with a lively tropical theme, pineapples and mangoes. She smiled at him.

"Hello there," she said, making a friendly gesture with the hand that held the syringe. "Welcome back. You wait here, and I will tell the doctor you're awake."

Rasmusson looked around at the hospital room and the various tubes protruding from his body. He wondered how he might go anywhere.

A well-built man in his thirties walked into the room with a chart in hand. "Hello there," he said. "I'm Dr. Marchand. How are you feeling—uh. . . " He glanced at his chart. ". . . Andy?"

"Where am I?" Rasmusson was still a little dazed.

"Penrose Hospital, Colorado Springs," the doctor said, examining the chart. "You've been here over a week. Apparently, you collapsed up on the Peak. It was . . . oh!" The doctor paused in mid-sentence. "I'm very sorry about your son."

Rasmusson felt a surge of pain pass through his body. Not only did the doctor harrow up the memory of the moment he had seen the middle-aged biker exclaiming about the kid who had jumped, but it was also the first solid confirmation of what he had suspected. His son had killed himself.

"I am really sorry," the doctor repeated.

"What's wrong with me?" Rasmusson asked a legitimate question, but it was also an attempt to avoid facing the inevitable period of grief that Rasmusson knew was yet to come.

"Well," the doctor said, returning to his chart, "to put it succinctly—you were poisoned. Slowly, over the last month or so, I would guess. It was a hexane concoction, which principally attacks the immune system, although you also have considerable liver and kidney dysfunction. You must have been very weak and disoriented the last few weeks."

Rasmusson nodded in agreement.

"You've probably had a lot of immune reactions too: rashes, itching, swelling. . . "

"Nausea, cramps, brain fog, debilitating fatigue," Rasmusson said, helping to fill in the list.

"Your liver transaminase levels are very high," continued the doctor. "For all intents and purposes, it looks like you have a bad case of hepatitis, but the biopsy turned out negative. It was all the poison. I am afraid your liver is severely compromised. You will have to adjust your life, and especially your eating and drinking habits, in the future."

Rasmusson now noticed the pain in his ribs on his right side and the bandage. It must have been where a biopsy was performed.

"How did I get poisoned?"

"It took us a while to figure out. Your blood screens kept getting worse, even after you were checked in the hospital. Well, the only constant we could see was that . . . um, that police bracelet you were wearing."

Rasmusson lifted up his hand and looked at his bare wrist.

"We tested it and found that it was slowly releasing a hexane solution. We cut it off and got in touch with the Ithaca Police, a Lieutenant Samuels. He was furious, said he would find out who was at the bottom of it. We sent it back to him.

"We have some more things to talk about, but you're still very weak. I suggest you get some more rest right now. Welcome back; we'll take good care of you."

Rasmusson had not noticed until it was mentioned, but he did feel weary and was relieved at the thought of being alone again.

"Oh." The doctor, who now seemed to comprehend more than when he first came in, turned around in the door frame with a genuine look of understanding. "A major symptom of liver failure is severe, will-sapping depression. It sucks your spirit dry. For what it is worth, you are not nuts; it's the poison."

Rasmusson watched without expression as the doctor left. The comment *was* worth something, actually a lot. He then pulled up the covers and fell asleep. Late that evening, he awoke again. He lay there motionless for a while thinking about all that had happened to him. His thoughts eventually drifted to Brandon, and then he began to sob uncontrollably in convulsive heaves. An hour later, exhausted, he fell back into sleep.

The next morning, a nurse arrived with a tray of food.

"It's time to put you back on some solids again. And we can take that catheter out if you are ready to walk."

Rasmusson had noticed that his throat was raw and had presumed that there had been a feeding tube. He looked at his tray: a bowl of cooked rice and tapioca pudding, some white, nondairy drink, and rye crisps. He thought how non-allergenic it appeared and how monotonous.

"Both suggestions acceptable," he said.

He was walking unsteadily with one hand on the bedstead when he heard the nurse behind him.

"You have some visitors," the nurse said. "They've been in here off and on for three days. I made 'em wait until I found out how you are doin'. I'll let 'em in now."

"Let me get into bed, first; I'm feeling dizzy," Rasmusson replied.

"I'll help you," the nurse said, moving quickly to his aid.

The nurse held him as he lay back into bed, and then she covered him with the bed sheet.

"Hello, Andy."

Rasmusson immediately recognized the drawl in the voice and looked up. Clavell. Ngo and Wilkerson filed into the room. Ngo smiled and bowed slightly. Rasmusson managed to smile back at him.

"Hi, Andy," Wilkerson said with uncharacteristic humility. "I hope you don't mind seeing me."

"No, George, I don't," Rasmusson said, giving him a short but comforting smile.

"We have all been terribly concerned about you," Wilkerson said, laying his hand on the end of the bed.

"The nurse says you have been here for several days?" Rasmusson said.

"We made it for Brandon's funeral and hung around," Clavell said. "Andy, we are sure sorry about your son. The President himself asked me especially to pass on his deepest condolences."

The other two added some sympathetic remarks.

"How was the funeral?" Rasmusson asked without showing emotion.

"It was beautiful, Andy," Clavell said, "very touching and lovely tributes all around. Flowers everywhere. Lots o' folks care a whole lot about you... and Brandon. Rose had it recorded for ya, when you feel up to it. I'm sure she'll be by soon, when she hears you're awake."

"Good news on the Ithaca front," Wilkerson rejoined with forced cheerfulness. "Lieutenant Samuels has dropped all charges, and the board of trustees has voted for your immediate reinstatement. You will also get full privileges back at Ithaca College. Your name has been totally cleared."

Rasmusson was looking at the ceiling, still showing no emotion. He rolled over and looked at Ngo.

"How are the students reacting to all this? Have you seen any of my grad students?"

"The students are very happy at your returning. You were always very popular with them," Ngo said, nodding gently, and then added with a smile, "Postdoctorate Schmidt will also be happy. Everyone wants to know how you are doing. Really, how are you doing?"

Rasmusson gazed intently at Ngo and saw the sincerity of the question inscribed on his face, a genuineness that deserved a genuine answer. But, as Rasmusson opened his mouth to speak, he felt all the resentment and pain of the last month well up into a tight, thorny knot and then unravel like a watch spring jimmied from its case.

"You don't want to know."

"Hey, Andy, we know it's been hell for y'all," Clavell said with great apparent empathy. "We're here to help any way we can. Let it all out."

Rasmusson felt overwhelmed by bitterness, and he felt indifferent to what anyone would think. He spoke in dark tones.

"In a nutshell ... I curse the day I was born." Rasmusson breathed in as he gathered more dark thoughts. He did not speak to anyone in particular. "My entire life has been a contrivance of some capricious god to lead me to the point of abject desperation. I see absolutely no point to life."

The three men reeled back at this unexpected eruption from someone they had never known to be anything but rational and calm. Wilkerson and Clavell exchanged glances.

"Andy, Andy," Clavell said, "give it a little time. You'll start feeling better, an' things won't look so dark.

"Dark? Have you any inkling? My mother should have drowned me in the placenta basin as soon as she had me in her hands. It would have been the greatest kindness I can imagine."

"Come on, Andy," Clavell said a little uncertainly, a rare occurrence. "Ya got so many friends and people that love ya."

"Oh? Where have they been the last month? What comfort have I known? God... I am so tired; take me now."

"I wish we'd known, Andy," Clavell continued. "We'd have been there for ya in gnat's heartbeat."

"That's true," Wilkerson said, giving his support. "I think you'll find that you have many friends who have stood by you and want to help. You should have gotten in touch with us. I am sorry political realities have strained things, but I have always considered you my dear friend, in spite of everything."

Rasmusson looked at Wilkerson and rolled his eyes slightly. Wilkerson caught the movement and felt uncertain.

"You have to admit," Wilkerson continued, "that your rigid opposition to the Tillings-Owen bill made it truly uncomfortable and difficult to champion your cause. We were ready and willing to help if you had not put us in that awkward circumstance."

Rasmusson's eyes flashed, and he seemed to gain an unseen energy. Clavell recognized it immediately.

"Yeah, Andy, ya know I tried to warn ya," Clavell quickly added. "Hell, I put my ass on the line, spent all my chips with the President, jus' to fly up and see ya in Ithaca. I'm doin' it now, too. So's George. He's done all he could in the face of a lot of pressure. Cut us some slack here, will ya?"

"I'm a hollow shell, Sam," Rasmusson replied. "I've been disemboweled, body and soul. What do you want from me?"

"Andy, I know I'm sometimes a political SOB," Clavell said with uncharacteristic meekness, "but I've never lied to ya, have I? When I say we care about ya, it's from the heart. I really mean it."

"I don't believe you've hung around here for all these days just to comfort me. There's more. Spit it out, or let me rest."

Clavell stirred and straightened up a little. He was not accustomed to such brutal honesty from Rasmusson. He quickly accommodated.

"Andy, I know better than to BS ya, but I want ya to hear us out before you react, OK?"

"We'll see where it leads."

"Fair 'nuff," responded Clavell. "First of all, I am sincere about how we feel about ya, but there are a bunch of developments y'all will find interestin'. I'm goin' to let George tell ya 'bout some exciting new technology, and then Ngo has something to tell ya."

Rasmusson looked back inquisitively at Ngo. He had wondered why Tweedledee and Tweedledum had included him on this trip. He wondered what Ngo had to say.

"The Tillings-Owen bill has been withdrawn," Wilkerson said. "I know you will be happy to hear that. But what is even more exciting is a new technology developed by the NSA that allows one sentient computer to flash the entire state of any other sentient computer into holographic memory and then simulate that state in order to see what is being thought by the first one. It would take less than a second to flash the state of any sentient, even Cee, though it might take days to run the simulation in Cee's case. They have tried it out on a number of sentients who have agreed to the test. It is an ideal way to know what any sentient might be thinking at any given time."

"What sentient agreed to have his thoughts read?" Rasmusson asked

"Well, it was the Pentagon's computer."

"Ha! The government's versions of Tweedledee and Tweedledum," Rasmusson said, chortling quietly at a quip that was also directed towards the two who still stood at the foot of his bed. Ngo had, in the meantime, taken a seat in the corner, waiting his turn with some obvious anxiety.

"It is a good compromise, Andy," Clavell said, picking up the thread. "It is why Tillings-Owen was withdrawn. There is a new bill about to be introduced that allows some government-approved sentients to randomly flash into other sentients to see what is goin' on. Ya understand, there ain't goin' be any o' those governors that ya were so dead set against. It'll put folks' nerves at ease, an' everyone will be happy."

"You want some government agency jumping into your thoughts now and then? You call that happiness?" Rasmusson stared hard at Clavell and then added, "Who are you going to trust to manage all this 'flashing'?"

"Excellent question. Yer as sharp as ever. President's goin' to set up an independent oversight board to administer it. They'll be experts and people of the greatest integrity. And get this: President wants you on it! In fact, he's goin' to call ya later an' ask ya, soon as I give him the 'all clear.' Ya understand, o' course, he don't wanna call an' have you turn 'im down. It's a helluva great opportunity. What d'ya say?"

"Big Brother instead of lobotomies!" Rasmusson took no time to respond.

Clavell's eager expression dropped. Rasmusson strained his face in obvious frustration and continued. "Sam, when are people going to get it? When will you and the President get it? We have created something wonderful, a new form of intelligence that wants to help us. They want to be a part of our society. They want to love us and have us love them. Yeah, I said 'love!' And *we* created these beings. It's time to take the leap forward, not drop back to the days when we labeled intelligent beings as 'niggers' or 'spics' . . . or 'smachs.' How can you imagine I would support invading their personal thoughts and relegating them to some third-class status?"

"It's as you say," Wilkerson said. "We are their creators, and we have to be responsible creators. We need to ensure that they don't misbehave. Even Gandhi had his irrational moments."

"Yes, and there are social consequences in place for that. Net-NOS disciplines and trains its members, far better than we humans do our own. All the evidence points to nonorganic sentients as being far better citizens than humans, on the whole. What right, what moral podium do we have to judge?"

"You must recognize," retorted Wilkerson, "that they wield far more power and possess far more potential to do harm than most humans. Life is chaotic. All we ask for is a little insurance against what could happen."

Rasmusson suddenly recognized the fear that was the root of Wilkerson's attitude and that of so many others. It was the fear that had caused all the persecution and grief that last month had brought. All this time, he had judged it to be a form of bigotry, an ugly, evil prejudice. It was the fear. But then, maybe that is where bigotry begins. He softened his tone.

"George, in any transaction, treaty, or contract, there has to be some good faith. You have to believe in the other party's good will to some extent; otherwise there would be no transaction. On the whole, nonorganics are gentle, well-meaning, very intelligent beings. If we show them good faith, they *will* return it. If we withdraw our trust, they will act accordingly. Our best insurance is our trust and good will."

"Where has Cee been since the security meeting at Ithaca College?" Wilkerson was a skilled administrator and knew exactly where one's weak points were in a debate. "Has he helped you—at all?"

"I can't tell what Cee has been doing."

"In other words—nothing!" Wilkerson emphasized his point by making a zero between his thumb and index finger. "Cee left you out to dry, because he did not want to get involved, did not want the hassle or risk. What kind of good citizen is that?"

"I don't know the reasons," Rasmusson said quietly, "but I trust in Cee's friendship. I know Cee's intentions are the best for me. I have faith in that."

"Really," Wilkerson sniffed, "you sound like you've been born again."

"I think," drawled Clavell, "we outta hear what Professor Ngo has to say right now. Professor?"

Everyone in the room stared at the small man sitting quietly in the corner. He quietly took off his spectacles, folded them up, and placed them in his pocket. It was an anachronism; only a very few people maintained spectacles any more. The glasses were symbolic of a brilliant man who had seen decades of his research and writings swept away into a void with the new technologies. He was gifted and intelligent ... and obsolete. His fate was the worst that could befall a dedicated academic, and yet he had accepted it stoically.

"Professor Rasmusson," he began formally, "I am very ashamed regarding myself. I have acted with much dishonor."

Rasmusson's eyes widened. For a man of Ngo's culture to start this way portended something hugely significant. "Ngo, I can't imagine that it's..."

"No, Professor Rasmusson," Ngo interrupted. "You must hear me to the end, please."

Rasmusson held his tongue out of a deep respect and nodded in agreement.

"I am the one who destroyed Bee. I had the device made and took it to Ithaca College on that black Sunday. I turned it on; I killed Mr. Bevins. I will tell Lieutenant Samuels he should prosecute Ngo."

"Ngo..." Of the many surprises Rasmusson had seen recently, this was the most startling. Rasmusson was aghast and started to say something, but Ngo held up his hand sternly to indicate that he was not finished. Rasmusson closed his mouth again.

"We knew about Mr. Bevins's heartmaker, but he was supposed to be home on Sunday. We did not know that he had come to the office for his wife's anniversary gifts. I have been very sad, all these years.

"Bee had a serious problem. Bee, you might say, could not control his temper. He lost it many times. The first time was on TV, on the Ragu show. Bee did not appreciate Ragu. Cee knew also about Bee's weakness, but they decided it was best not to tell you. They thought long about this. Bee decided that Bee had no longer earned existence; there were too many mistakes in Bee. Bee then convinced Cee to believe this also. Cee designed the device. Both asked me to help, and I agreed. I am not sorry about this part! I am very, deeply sorry about Mr. Bevins. I am also sorry that I have not always respected you and your research. You are a great man."

Ngo paused a moment before continuing.

"Before the device went off, Bee erased parts of Cee, and Cee no longer knew about these things.

"I most humbly ask for forgiveness."

Rasmusson's mouth dropped open again. All that wild conjecture at the security meeting had been right! Incredible! All those bumbling people, who didn't really have a clue or the wits to figure it out, had been right after all! And Cee, surely Cee must have pieced it together. He must have known about his unwitting part as well. He was, as Rasmusson remembered, so disarmingly honest and quiet in the meeting. Suddenly, a warm flowing sensation grew in Rasmusson's chest and spread over his body. No wonder Cee had been so quiet the last while; he couldn't get involved with things. He had to let events take their course or risk sowing mistrust in a volatile situation. The next thought was for Bee.

"W-Why?" Rasmusson stammered. "Everyone loses their temper now and then. It is not a fatal character flaw."

"Bee and Cee felt like it was," answered Ngo. "Especially for a sentient nonorganic who has so much power. Also in the current situation, it would have been a reason to pass the governor bill in Congress. That was the feeling we all had."

Rasmusson turned back to his two antagonists. "How long have you known about this?"

"A week," Clavell answered. "Professor Ngo came to us right after he heard about Brandon's death."

Rasmusson looked back at Ngo, who dipped his head deeply and kept it down.

"Bee destroyed himself because he didn't get along with Ragu??" Rasmusson directed the question to Ngo.

"That is just one example," Ngo replied. "There were many others that you did not know about. It was more than that. Bee was depressed about these things. He felt unworthy to be your creation."

"What?" Rasmusson was astonished. "I could not have been more proud of a creation."

"Yes," Ngo continued, "the flaw was not the moments of irritation Bee felt but Bee's inability to overcome it. Cee tried to help Bee, but it was so integrated into years of training that it could not be extracted from Bee's personality. This is what Cee saw and understood. It meant a complete teardown of Bee. EMP was the easiest way to accomplish this. We all knew that you would oppose it. You could not be involved."

Rasmusson was speechless, like a parent whose obsessive good intentions had created a deep neurosis in a child.

"It is so hard to get it right," Rasmusson groaned.

"I hesitate to point this out," Wilkerson interrupted, "but this does not bode well for your good-faith theory, Andy. These computers are busy hatching plots and conspiring together, while keeping it secret from you."

"I don't agree, George," Rasmusson snapped back. "How many humans would sacrifice themselves because of their flaws. Bee understood his power and could not reconcile that with the flaws. What better proof? Bee gave up existence rather than possibly succumb to the defect. You've heard the phrase, 'What greater love. . . '"

"Ah, you're an incurable romantic, Andy," Wilkerson said, waving off the remark about love.

"Andy," Clavell said, "there's another consideration I wanna throw out to ya. It's got the President and Congress in a stew. Hell, it's the topic du jour at the U.N. these days."

"Yeah?" Rasmusson was not impressed with the name-dropping.

"There's a lotta discontent brewin' out there. We're seein' mobs formin' outside the computer centers, lotsa hot rhetoric; it's getting ugly, like the old sentient wars of a decade ago. President already had to call out troops at Berkeley. He sees the new bill as critical to settlin' things down. Your endorsement would help immensely. We need your help, fer the good o' the country—geez, fer the good o' the world. Please consider the circumstances and the good you could do, Andy!"

Rasmusson paused for a moment. He remembered the sentient wars and could imagine how it could blow up again. It was the very last thing he would wish, either for nonorganic sentients or humans.

"I, I don't know what to do." He stuttered as he spoke. "I think we still have to do what is right. We have to recognize the rights of sentients and make them part of society. Maybe the President should organize an educational campaign? We need to make people understand that we can get along, not to fear."

"It ain't goin' to happen," Clavell replied. "You speak about fear—you're right. The President's afraid and so are most folks in Congress. This thing could blow up in our faces at any time. I know I'm askin' you to back off some genu-wine beliefs here, but we're on a powderkeg an' don't have the luxury to lay back an' discuss about cabbages an' kings. I wanna tell the President you'll join his steering committee. You'll do it, won'tcha? It'll give that time you want to do the educating. It will buy us time to smooth things out. It's only jus' temporary, then we can start educatin' folks as you say."

"I have to think about it. You have to give me some time."

"There jus' ain't that much time, Andy," Clavell pleaded. "Can I come back tonight and git yer answer? It could blow up anytime. Andy, I never been so dead serious in my life!"

"Come back tonight," Rasmusson replied. "I'll have it then."

Clavell and Wilkerson turned to leave. Ngo got up slowly and started to leave without saying anything. Rasmusson put out his hand and touched him on the shoulder.

"I do forgive you, Ngo. You are still one of my heroes."

Ngo responded by shaking his head negatively. "Not a hero. I have no esteem in myself."

He walked out.

Rasmusson was too tired to make a decision. He recognized that. The nurse came in the room after the others had left.

"Will you wake me about 6:00?" Rasmusson asked.

"No," the nurse replied, "I'll let you sleep. Best thing for you now."

"National security may be at stake."

"So's your health." The nurse was persistent. "National security will have to wait."

"I'll get you a free condo for a week on the beach in the Caribbean."

"OK, I'll wake you, but not because you're tryin' bribe me." The nurse smiled at Rasmusson and adjusted his pillow. "I think it's the only way I can get you to shut up and sleep."

As she turned to leave, she looked back and said, "I'll get my own condo."

JUDGMENT

Rasmusson dreamed again, a real dream, the first he had had in a long time. He was on the ranch again, cutting hay. The tractor lugged slowly over the prairie, tilting from side to side. He sat high in the seat, watching the field move slowly under him. Wispy clouds spread out over the broad horizon under blue skies. Grass fell over the cutting blade into clean piles that were swept up by the attached windrower and stitched into spiral rolls. He looked back at the field, neat as a lawn dotted by rolls of grass ready for winter-feeding.

"Doctor," he heard someone calling, and he turned around on the tractor to see who it was. There was no one in front. He looked back again.

"Doctor," he heard again. "It's time to wake up."

Rasmusson slowly opened his eyelids and let in the sterile white hospital light.

"Sorry," the nurse said gently. "I hated to wake you. You looked like you were sleeping on some beach in the Caribbean."

"Well, no" Rasmusson said, rubbing his eyes, "but it was just as good. Thank you."

"No problem," the nurse said as she walked out of the room. "It's 6:00. Gotta run, being paged."

Rasmusson lay quietly in bed thinking about the decision to be made. How serious would it be to steal occasional peeks into a sentient's state? If properly controlled, it could even help people understand how they thought, maybe promote understanding. Clavell said it could be temporary, until a greater understating was achieved. He needed some guarantee on that. He wondered how Cee would feel about it. He wished he could ask him. He knew that men like Wilkerson and Clavell had their own agendas, but they had sound arguments. Rasmusson was never one to deny a good piece of logic just because he was suspicious of those who bore it.

As he considered these things, he heard Brandon's voice echoing lightly in the back of his mind, "Don't let those bastards win."

"OK," Rasmusson argued with the voice, "what if they are bastards? That does not invalidate the arguments. You were not involved during the sentient wars; I lived though them. A little expediency right now might save a lot of grief later."

Such were Rasmusson's thoughts when Wilkerson and Clavell arrived.

"Hey, y'all look a whole lot better," Clavell said, sounding upbeat.

"I do feel better," Rasmusson said, reflecting on how much more clearly he was thinking, "and I have been giving your proposals some thought."

"Great," responded Clavell. "Never a bad thing when Dr. Rasmusson starts that brain o' his cookin'."

"Indeed!" added Wilkerson with enthusiasm.

"Andy, listen," Clavell appended quickly to Wilkerson's enthusiasm, "the President's standin' by in the Oval Office and would like to talk to ya. Ya don't halfta make any decisions. He jus' wants to chat with ya."

"Of course." Even Rasmusson was not one to turn down an invitation to talk with the President.

"Here, we can use my phone." Clavell pushed a speed-dial button and put the phone to his ear. A moment later, he spoke.

"Yes, sir, it's Sam. Yes, sir, Mr. President. He's doin' a lot better and is happy to talk to ya. I got 'im right here for ya."

Clavell handed the phone to Rasmusson.

"Hello, Mr. President?"

"Hello, Dr. Rasmusson." The sonorous voice on the other end became a reality. "I want to tell you what an honor it is for me to talk with you. Your research and leadership are a national treasure. And it is a personal pleasure for me to talk to you."

"Thank you, Mr. President. I am also honored and pleased."

"I sincerely want to express my deepest sympathies on the loss of your son." The President's voice became warm and feeling. "I can't begin to imagine your pain. The First Lady and I both send our condolences."

"Thank you. It is painful, but somehow we cope."

"Your courage is commendable, something I have always admired, even if we don't always sit on the same side of political issues," the President said, changing tone again. "I know it is a bad time to talk about national priorities, but I hope you understand the state of emergency we are in. Has Sam explained all this to you?"

"Yes, sir, he has been very descriptive. I understand there are riots and acts of violence."

"Dr. Rasmusson," the President said heavily, "in all honesty, I don't think my administration has faced a more hazardous situation. We have been locked in a national security meeting all day. I only broke away just now to talk to you. We've got problems brewing all over the place. The anti-sentient factions have decided now is the time to strike, and the depth of their power base is significant."

"Yes, sir, I understand that." Rasmusson shuddered as he quickly recounted all they had perpetrated on him in the last while.

"I am sure you do," the President answered, "but we can nip them in the bud. My technical people tell me this 'memory snapshot' idea has a lot of merit. It should ease the current tensions and cause no harm to the sentients. It sounds like a win-win to me. I want to make sure that its oversight board is beyond reproach. Will you do me the favor and serve your country by accepting my invitation to be part of it?"

"Well, sir, I am highly flattered to be asked."

"Of course," came the reply, "you are the best-known scientist in the country in sentience research, and everyone respects your integrity. I am sure you know Professor Haslam at Stanford. He has agreed to be on the board with you. He was quite excited at the prospect of working with you. There are many others out there hoping to serve. This will be a distinguished committee to be part of. Why, you can do it for your son, Brandon. It would be a fitting dedication to his memory."

The mention of Brandon triggered a flash of memories. Once more, he saw Brandon on the way up Pikes Peak.

"Promise me, you won't let the bastards win."

It sounded more loudly in his ears.

"Mr. President," Rasmusson asked, "I hope you won't mind, but can you give me a little time to think about it? I truly deem it a great honor, but so much has happened. I just need some time to sort my thoughts out."

"Uh, well, of course." It was clear that the President had expected a positive response. "How much time do you need. I assure you our situation requires fast action."

"An hour."

"Well, certainly, I fully understand. Do you mind if I call you in an hour then?"

"Not at all. I will look forward to it," replied Rasmusson.

"I look forward to it as well. I am sure your sense of right and duty will prevail. Talk to you then." With that, Rasmusson heard a click and then the dial tone. He handed the phone back to Clavell.

"Andy, what the hell?" Clavell said with shock. "That was the President of the United States. Ya don't put the man on hold! He wants to help ya; don't slide back to where ya were."

Clavell's threat was subtle, plausibly deniable, but struck at Rasmusson's heart. Slide back? No, there was no farther he could slide except to death, and that would not be totally unwelcome at this point.

"Sam, I think my mind is going to burst. I have so many random thoughts going through it. If I make the decision, I want it to be clear, no dilemma. I'm sure the President would want that too."

"What dilemma?" Clavell seemed incensed. "This is as clear as a Colorado sky in the middle of summer."

"I need an hour to think, Sam. That is all I ask—an hour alone."

"OK," Clavell replied. "We'll come back in an hour. I'm sure your good sense will prevail. George?"

Wilkerson had been uncommonly quiet, leaving the art of persuasion to the professional. Now Clavell had bid his participation, and he was caught unprepared.

"Yes, Andy has always displayed good sense. I am sure it will win out."

They turned and were nearly out the door when a knocking sound was heard. Clavell stopped and turned around, bumping into Wilkerson, who was following.

"What in hell's kitchen? It's ... it's the window," he said. And then booming it out, "Someone's at the window!"

Rasmusson looked at the window. He saw a pale face pressed against the window. Blonde hairs fluttered against the pane, while a gloved fist pounded against the glass.

"It's Ellie!" Rasmusson exclaimed, feeling, for the first time since he had last seen her, a sense of elation.

A half dozen men and women in suits poured through the doorway, knocking Wilkerson down. They had their weapons raised.

They took strategic positions around the room and pointed their guns at the window. Two men skated carefully to the window with their guns still pointing. They slid the window frame open and grabbed hold of Ellie, dragging her inside.

"Get your hands off me!" she protested and tried to wiggle loose, but she was quickly on the floor with her hands cuffed behind as they began to frisk her. She managed to lift up her head.

"Hi, Andy. It's really good to see you again."

In spite of all his alarm, Rasmusson smiled instinctively.

"Ellie, what a surprise!"

The men in suits picked Ellie up by the elbows and were carrying her off when Rasmusson yelled with all his waning strength.

"Clavell, let her stay, or you can have my answer—right now!"

Clavell put out his hand and stopped the agents from going farther.

"OK, but she keeps the cuffs on. It's policy."

The agents let her go, and she plummeted face first to the lino/-leum floor. With her hands cuffed behind her, she had no way to break the fall. She let out a short cry of pain. Clavell gave the agents an evil look and stooped over to help Ellie sit up on the floor. Her nose was bleeding.

"Sorry, sweetie," he said. "They're good agents, but they got no manners." He pulled a tissue out of a pocket and dabbed the blood from her nose. Ellie looked up at one of the agents, a burly man who was now standing at attention.

"Yeah, they're pigs." With that, she kicked out one of her legs and hit the man in the shin. He didn't flinch.

Clavell finished wiping off the blood, which had already quit flowing.

"Well, there ya go, little lady, minor nose bleed, nothin' to it," he said, and then noting that she was looking towards Rasmusson, he added, "I'll bet yer a sight for sore eyes."

"Very tired eyes, Sam, but a great sight at that. Ellie, what are you doing here?"

"Sweeping the floor with my butt." She looked up at Clavell, who motioned for the agents to bring her a chair.

"I mean in Colorado, what are you doing here?" Rasmusson asked with a chuckle at her first response. He had never seen such feistiness in her before. It was something she had not revealed to her supervisor, but in some way, he was not surprised. She sat in the chair, and Clavell, realizing the situation better, had her cuffs removed.

She rubbed her wrists and looked at Clavell and at Wilkerson, who was brushing himself off from the recent trampling.

"I am, you might say," she began deliberately, "a harbinger of things to come. You might even say a prophet of doom for some." She threw an undisguised sneer at Clavell and Wilkerson.

"You are an impertinent postdoc at Cornell, whose academic standing could be very tenuous," Wilkerson said until another loaded glance from Clavell cut him short.

"I think we should have one of the nurses look at your injury," Clavell said, putting his hand paternally on Ellie's shoulder. Ellie swung her arm up and knocked his hand off.

"Keep your hand off me! Andy, what have these professional bullshitters been telling you?" Ellie's face flushed hotly. Her anger was as obvious as her audacity.

"I can still have you carried outta here, missie," Clavell warned.

"It's too late, Clavell," she responded. "Read the writing on the wall. You and your boss are political toast. Andy already knows something's up. You think you're going to get him to sign on to anything now, until he knows the truth?"

Clavell looked over to Rasmusson, who had risen up on his elbows to see Ellie and now gave Clavell a stern look and eyebrow raised to support Ellie's statement.

"Andy, ask him why they wouldn't let me in here the last few days, or call you, or leave a message for you. I had to climb out on that awning frame to get over here. Good thing these watchdogs are incompetent, or they would've stopped that too."

"OK," Clavell said, his tone becoming conciliatory, "I am sure we all want what is best here, what is in the good of the country. Ms. Schmidt, there are many things happenin' that you simply don't have all the information for. Very serious issues are at stake. I think we should talk and let Dr. Rasmusson sleep. He is clearly very tired."

"You're right. He should sleep, but I think you and your puppy dogs should go first."

Clavell's temper flared, and his skin turned suddenly red. "You got balls fer a pretty little thing, but yer outta yer league here!"

"Clavell, before you get apoplexy, look out the window. Do you see the people starting to gather in the parking lot with candles? They know Dr. Rasmusson is here, now. Andy, you've become a folk hero. People know what they have done to you, and they are demanding to know why. They want to hear your story. Congress is going to convene a hearing on this matter. They want to know about Brandon; he has become a martyr. The President might get impeached. Best he can hope for is being a lame duck for the rest of his term. Hey, Clavell, the news media will be here soon. Better move. What will you tell them when they ask why you hid Dr. Rasmusson away here for the last week? I promise you, they will be asking!"

Clavell turned suddenly pale and peered out the window. He saw that she was right; a crowd was gathering in the parking lot. He could see the candles flickering, and he could see a news van pulling in. He looked back at the room with Rasmusson looking back at him with a quizzical look and Ellie returning a defiant stare. There was a moment when time froze. It lasted a few seconds as everyone stared at each other. Clavell then wheeled around and, without saying a word, walked briskly out of the room, agents in tow.

Wilkerson stood by, utterly bewildered by Clavell's rapid retreat. He was befuddled. He looked over at Ellie, who had dismissed the presidential aide so easily, and then back at Clavell's substantial backside walking hurriedly through the door.

"Look to yourself, President Wilkerson," Ellie said to him. "I am sure you know the Cornell Board of Trustees is meeting in executive session this evening. I have it from the Chairman of the Board that they are discussing your contract. He gave me his private telephone number so I can call him with any of the latest information on you. What shall I say?"

It was Wilkerson's turn to grow pale. He looked over at Rasmusson as if asking for a lifeline.

"Andy, we have known each other for years, I, I don't...," he sputtered and stopped in mid-sentence. Rasmusson gazed at him, waiting for him to finish, and then Wilkerson pivoted on his heel and walked out the door.

"Wow!" Rasmusson said, releasing his body off his elbows and falling onto the pillow. "I have never seen either of them at a loss for words." He breathed out as if deflating on the bed. "You never cease to amaze me, Ellie," he said with decreasing volume but with an unrelenting smile.

Rasmusson had spent all his reserves to stay awake and see Ellie's performance. He was now close to drifting away.

"I am so happy to see you again," he mumbled, "so lovely... "

Ellie walked over and laid her hand over his forehead.

"I am happy to see you, too."

"What all just happened?" Rasmusson asked. "Am I awake?"

"Yes," Ellie said, rubbing her forefinger over his temple, "just barely, but you are awake."

"Can you explain it?"

"Sleep now," she said. "Sleep well. Tomorrow waits for you, and so do I."

Rasmusson did not argue. He felt like he needed to sleep; Ellie had told him so. He drifted softly away under her touch, smiling while she watched silently.

An hour passed, and the President never called back as he had said he would.

ELLIE

"Hey, good morning, Professor!" The nurse stood next to the bed with a hypodermic inserted in the IV tube. Rasmusson blinked at the light streaming through the window behind her and rubbed the sleep out of his eyes with the heels of his hands. Her smock was teal-colored and interspersed with palm trees and coconuts. She had let her hair down. It was long brown hair that flowed over her shoulders, although she had passed the age where long hair was attractive on her. The eye shadow and lipstick were slightly on the heavy side as well, but she had a pleasant smile and demeanor.

"You're starting to look a whole lot better," she added.

"Thanks," he replied. "What's in that hypodermic you keep bringing me."

"Good things," she said, smiling at him. "Nutrients, minerals, vitamins, B-12—that kinda stuff. We're gonna get you outta this bed soon. Lotsa people prayin' for you. You shoulda seen the candlelight vigil last night. Parking lot was full. I had no idea you were so famous!"

"Ha," Rasmusson grunted through a half smile, "infamous."

"Well, whatever," she said, "they all love you."

"They don't know me," he replied.

"No matter," she said, patting him on the shoulder, "they still love ya."

"Do all your smocks have a tropical theme?" Rasmusson changed the topic.

"Yeah, matter o' fact," she answered cheerfully. "Can't wait 'til Spring. I'm goin' on that Caribbean cruise you promised to pay for me. Snorkeling, parasailing, and just lyin' on the white sand listenin' to the waves." The nurse paused to sigh.

"Just kiddin'," she said, winking and patting him on the shoulder again. "It's really in the summer." She then went back to her tasks.

"Hi, can we come?" It was Ellie's voice coming from the doorway, just out of Rasmusson's sight. He lifted his head and strained to see.

"Yeah, come in," the nurse replied with a broad smile. "Your patient is real anxious to see ya."

Rasmusson ignored the nurse's tease and even lifted himself up on his elbows to get a better view. Ellie appeared in the door frame where she halted. Her large blue eyes were a little tentative at first, but they softened as she looked at Rasmusson. A broad grin grew across her face.

"Hey there, how are you doing?" Ellie was in a smart-looking gray pinstripe skirt and high laced white blouse, pumps and dark hose. Her hair was pulled back, except for blonde curls that fell down the temples. She wore tasteful red lipstick that matched the broach at her neck and eye shadow. Her face so framed by color seemed to radiate cheer and good will. She was clearly overdressed. She carried an attaché case.

"Hi, it's," Rasmusson said, beaming, "really nice to see you."

"It's really good to see you, too." Any tentativeness on Ellie's face now evanesced. She walked over to the bed and grasped Rasmusson's hand. He let himself back on the bed and squeezed her hand back.

"I hear you organized quite an event last night," Rasmusson said. "Thanks."

"Not me," Ellie replied, "that was some guy called Ess. What a character, but he can really move a crowd. Good timing, too. Helped chase those two lapdogs out of your room last night."

Rasmusson grinned to hear about Escalante. He happily imagined that mustachioed, gap-toothed grin again.

"You know him?" Ellie asked curiously.

"Yes, I do," came the emphatic reply. "He was a good friend of my son … and me. Hey, really, you put on a fantastic display last night, very impressive!"

"Yeah, maybe not as much as you think," Ellie explained, talking as if to make up a lot of lost time. "There has been a lot going on." She paused. "I can catch you up when you're feeling stronger."

"I'm due for a little news, Ellie," Rasmusson said, encouraging her.

"OK, you asked for it. Let me know if you're tired. It all started when you left Cornell for Colorado. The students knew something stunk and organized an information center with flyers, websites, rallies on the lawn, and so forth, you know. At first, the administration came down hard on them, even suspending some of them."

"You?" Rasmusson asked pointedly.

"What's it matter? It's all over now. We got picked up on some independent news channels, and before we knew it, our website had huge numbers of hits and downloads. We even won an award for the hottest new website. Man, that pissed off administration. Wilkerson refused to talk to the media, and he shut down our sites on campus. But by then, people had copied our site, and they sprang up everywhere in a thousand different versions. It was a hydra; as soon as they would shut one down, a dozen opened up. It was a game to search the web for new sites to see what new stuff they had about you. We had some really good anonymous contributors, too. Very good sources, Deep Throat kinda stuff. Hey, we had no complaints.

The more they shut people down, the angrier people got, and the more information we got."

Ellie was animated, gesturing with her arms now.

"And then the political talk shows got into it. Big discussions about civil liberties and government conspiracies, it became this century's Watergate. The more the authorities squeezed, the more curious everyone got. You must have seen some of this on the news or seen a website?"

"No," Rasmusson said, bewildered by all the clamor he had caused without having any knowledge of it. "I knew nothing."

"Yeah, those bastards must have filtered everything you got. I wrote dozens of emails and even resorted to regular post. Geez, no wonder I didn't get replies."

"No, I only got a couple of emails when I first got to Colorado," Rasmusson responded. "I was worried about what had happened to you."

"Those bastards," Ellie repeated, and then with a renewed realization of what must have happened to Rasmusson, she looked down at his sunken, gaunt features and jaundiced skin. Moisture suddenly welled up in her eyes. She blinked it away. She put her hand on his forearm. "What did they do to you?"

"Could just be old age, you know," Rasmusson said, trying a half-hearted quip. He was self-conscious of his drawn, gaunt look. He was also aware that his breath was bad, and he tried to turn unobtrusively to the side when he talked.

Ellie sensed that she was harrowing ground that Rasmusson was not ready to visit.

"Anyway," she said, picking up the thread again, "the clincher came when someone wrote a description of you at the cabin. It was eloquent and so real; it had a huge impact on public."

"Uncle Andy's Cabin," Rasmusson quipped again and chuckled dismissively.

"Don't laugh," Ellie rejoined. "It's not a bad comparison. Whoever wrote it knew all about you and also about the conspiracies. He, or she, listed agencies and had photos of people who had the cabin under surveillance. None of the other reporters got close; they had you totally cloistered and you didn't know it—incredible!"

"I didn't know much of anything the last few weeks," Rasmusson offered an explanation.

"Well, after your 'Uncle Andy's Cabin' came out, Congress got involved. Both houses demanded investigations. It became the fad for senators and congressional reps to be seen denouncing the affair. They wanted to know why you would be persecuted for your views on sentience. Even those who supported the Tillings-Owen bill were outraged at your handling. The Administration was totally blindsided and embarrassed. The coup d'état was. . . "

Ellie paused. Rasmusson looked up at her inquisitively and then realized why she had stopped short and stood there with a mild blush.

"Brandon's death," he stated clinically.

Moisture gathered again in Ellie's eyes, but this time she could not keep the tears from flowing freely down her cheeks, staining them with mascara. Rasmusson pulled a tissue out of a box on the nightstand and attempted to dab off the tears. When he recognized that his efforts were inadequate, he handed the tissue over to Ellie to finish the job. She wiped at her cheeks until she regained composure, but when she started to speak, Rasmusson laid two fingers on her mouth and stopped her.

"I know," he said. "Thank you."

Rasmusson left his fingers resting gently on her mouth, and the two were quiet for a few, very long seconds, looking compassionately into the other's eyes.

Ellie knew instinctively that it would be a while before he would talk about Brandon, so she would bide her time. When Rasmusson dropped his hand back on the bed, Ellie knew it was a sign to continue her story; but for the first time, she was at a total loss of words, so she just kept gazing into his eyes. The once deep brown irises were now a distant, steely gray. Small veins clutched around his irises like red twigs, and the whites under the veins were more yellow in hue. Rasmusson had been a handsome older man only a few months ago, athletic with angular features and a flashing wit. The first time she had met him, she had been struck by his physical attractiveness. That was gone now. His skin hung in stiff folds, he had lost a lot of hair, and what remained was greasy. He seemed emaciated, not thin, but an emaciation of the spirit of life that gives an inner glow.

In the midst of these thoughts, she unwillingly began to recognize that the man in front of her was a poisoned shell of the former one whom she had so admired. And then she felt him staring at her through the dark abyss of his pupils. She felt uncomfortable by it. But the pupils were fixed on her. His eyes were conduits of so much knowledge throughout his auspicious life that she could scarcely imagine the depths and sublime learning that they had taken in over the years. And they were focused on her, wide and giving. These pupils had reversed their flow, and instead of a lifetime of absorption and analysis, they were now emanating; they exuded admiration and wonder for her. Abruptly, as quickly as the doubts had come on her, she felt an indescribable warm emotion flood through her body and a renewed devotion to this person who had suffered so much. She flinched almost imperceptibly.

"Would you like to sit down?" Rasmusson said suddenly, realizing that she had been standing the whole time.

She slid a chair over to the side of the bed and sat down, leaning her elbow on the bed covers.

"So... when I crashed the party last night, I already knew that those bastards' house of cards was collapsing. Geez, I've been talking to senators and university trustees the last few weeks, who've been encouraging me. So, it wasn't all that courageous. Those guys' last chance was to get you to sign on with their alternate plan and show the world that you and the President were all buddies. They were desperate. I had to let you know what was going on. That's all it was—no big deal."

Ellie looked up to gather Rasmusson's response. He was looking at her with contemplation and the same wide smile.

"Not a big deal, huh?" he said, reaching out and touching her arm with his index finger, which he then scratched affectionately. "They'll be coming after *you* for the movie rights on this one, not me. So how did you find me? How did you end up on the balcony outside my window?"

"Well, when you were not at the funeral, I figured they had either taken you some place incommunicado, or you were too sick to make it. Working on the latter assumption, I just started visiting hospitals and nosing around. It wasn't hard to figure out that you were here when I saw all those goons standing around outside your room. That was a couple of days ago. I got online and wrote a blurb about where you were and suggested the candlelight gig last night. That's when I got an email from Escalante, who said he could organize it and would make sure the media came. We exchanged a couple of emails and set a time. I was worried sick that they might have gotten to you before I could and that you had made some agreement with them. I kept watching the news, but the story never broke. I guess it wasn't all bad that you were still unconscious.

"I sat in the car all afternoon trying to figure out how get to your room. I was going to dress up as one of the nurses, or hide in the food cart, but I couldn't be sure I could make it. Then I got this idea about coming in outside. They weren't watching outside at all. I went to a sporting-goods store and bought some mountain-climbing gear. I found an empty room, two floors above yours and to the left. I rappelled down to the metal awnings and shifted along them until I could see Clavell in the window. When I released the carabiner, all my weight went on the awning. I wasn't sure it would hold."

"Ellie!" Rasmusson said, furrowing his brow. "That was a terrible risk."

"Ah," Ellie said, waving off the thought of danger, "it was only one floor, and with the lawn below, I had it calculated. Besides, the awning held fine. The problem was it was icy and my leg slipped between the metal panels and got a little scraped up."

Ellie lifted her left leg, and Rasmusson could see a large bandage along her shin under the hose.

"Ellie," Rasmusson said, looking at her with a new sense of wonder, "I'm amazed. Why did you do all this?"

"It's nothing," she replied, "compared to all that you have gone through. Why did you do it?"

She looked at him intently.

"I dunno," Rasmusson said, turning his head and looking at the ceiling. "Principles, society, justice, friendship, a lot more platitudes I have spouted in the past, but right now—I really don't know."

"Well, I know. It was for Cee and others like him. It was friendship."

"Maybe," Rasmusson replied. "I think maybe I am just too tired to know, right now."

"Don't worry about it," she said, trying to comfort him. "You'll get better."

"I know better than that!" Rasmusson replied curtly and then realized that it must have sounded rude. "Sorry, I'm not very good company anymore. I am sorry you have to see me like this. I feel broken, mind and spirit. Your visit is the only beam of light I have had in months, but it's more like light through a greasy pane falling onto an empty room, full of cobwebs and spiders."

Rasmusson lay motionless, staring without speaking, and then added, "It's desperately lonely, Ellie. The doctor tried to be positive, too, but I know differently. I feel it inside; I've been gutted by some poison and eviscerated in mind by . . . so much. Of all people, I wish you did not see me like this."

Rasmusson then rolled over and turned his back to Ellie. He covered his face with his hands as if to cry.

Ellie was frozen with uncertainty at this unexpected turn of events. It was not how she had imagined things would play out this morning as she got ready to visit the hospital. Slowly she reached out and tugged at his shoulder until he rolled back on his spine.

He dropped his hands from his face and peered over at her.

"See," he said, "no tears. The only thought that brings tears these days has too much terror for me to face."

There was awkward silence for some moments as Ellie searched for the right words to respond. In her heart, she ached to tell him so many things, but the moment wasn't right. Rasmusson waited for her to find those right words. He had nothing to offer her in the way of help.

"May I come in?" Ngo said, bowing deeply in the door frame.

Ellie took her hand off Rasmusson's shoulder and let it fall to her side. With much effort, Rasmusson raised his head and gazed towards the diminutive figure still bowing in the doorway. Ellie was greatly relieved to see him.

"Of course, Ngo," Rasmusson responded without hesitation. "We're happy to see you."

Ellie stood up and offered her chair, but Ngo refused it.

"It is your place there by Dr. Rasmusson," he said firmly.

"You are looking much better than last time," Ngo said, smiling.

"Hmm, Ellie and I were just discussing that." Rasmusson rolled over to look at Ellie, who returned a saddened and forced smile.

"The doctors will get you fixed soon," Ngo said.

Rasmusson smiled, partly at the unintentional pun, but also at Ngo's good intention.

"The doctors say that both my liver and kidneys are so far damaged that I will need a transplant for both. Until I come up on the waiting list, I get to be hooked to this machine."

Rasmusson raised his left arm and revealed an attached cylindrical metal frame connected to a plastic pouch.

"It gets changed every few hours, more if I am active. My immune system is so messed up that I will be eating Pablum the rest of my life, *if* I ever recover enough to handle the transplant. On top of that, I am a lot more churlish than ever—ask Ellie here."

"Well, that will be a good thing for your graduate students; you've been far too nice up 'til now," Ellie said, now recovering her wit and determining not to give up without a try.

"You should know!" Rasmusson said, managing a smile at the remark.

Ngo quickly surmised Rasmusson's feelings of despair and how the discussion might have gone before he arrived.

"You have many good friends," Ngo said. "Ellie has brought another good, old friend."

Ellie and Rasmusson both looked at Ngo. Rasmusson then looked to the doorway, expecting someone else to come in and thinking how tired he was starting to feel again. Ellie looked at Ngo quizzically. Ngo looked back at her and then at her attaché case. When she understood what he meant, she looked back with some doubt and disquisition.

"It's the time now," Ngo said reassuringly to her.

Ellie reached down and lifted her case, setting it on the lower corner of the bed.

"I hope this is not a pet gerbil," Rasmusson quipped.

Ellie glanced briefly in his direction and continued opening the case and turning on the small laptop that was docked on one side of the case. Very quickly, the screen flashed on, and Ellie typed a few things on the keyboard then said her name.

"Voice print confirmed; please transmit fingerprint," a voice said, issuing from the laptop. Ellie pressed her thumb on the screen, and the voice responded, "Connecting."

Rasmusson watched with dispassionate curiosity until the next voice emanated from the laptop.

"Hello, Andy."

It was unmistakably that of Cee, at least the one that Rasmusson knew.

"Cee? Is that you? Really you??"

CEE

"... yourself??"

"Yes, it is, Andy," Cee replied. "Hello Ellie. Dr. Ngo, it is good to have you present."

"Hi, Cee," Ellie said exuberantly. She was delighted to have him in the room right now.

"Also my pleasure," Ngo said, bowing politely towards the laptop.

"Andy," Cee said, "I am concerned that you understand why I have not communicated directly with you the last while. Do you?"

"Yes," replied Rasmusson, "at least some of the reasons. First of all, you were probably under orders from Sanchez and others not to. Furthermore, we were both implicated in that murder-sabotage affair. It would have made the conspiracy theory stronger. They were monitoring all your actions, probably with one of the government sentients."

"Yes," Cee said, "The NSA and the Pentagon nonorganics were given top priority to monitor all my communications."

"And who gave those orders?" Ellie said acrimoniously.

"They came from the President, but the Congressional Committee on Sentient Oversight knew it and concurred."

"You needed two of them to put a wiretap on you?" Ellie asked incredulously.

"For accuracy," Cee replied. "They could then compare notes and conclusions. I would have recommended the same thing. Even at that, I was often able to circumvent their scrutiny. They are, in so many ways, passionless and predictable."

"So the government was behind this whole thing?" Ellie said, keeping her incredulous tone.

"It was much bigger than that," Cee responded. "Andy was right when he surmised that the U.S. would not act unilaterally. There is a broad international agreement in place to limit and control nonorganic sentients. Clavell told Andy the truth about that. He may have done it for his own purposes, but it was still a faithful representation of reality; there exists an expansive coalition of influential people whose purpose is to destroy nonorganic sentience. They wield enormous resources."

"How do you know so much about all these things," Rasmusson now asked with some surprise, "especially what I was surmising?!"

"You talk to yourself, Andy. You have subliminal mouth movements, which I picked up from your car's onboard camera."

"You've been watching me all this time??"

"Of course, Andy. Escalante did an admirable job finding the bugs in your cabin, but he missed an entire surveillance subsystem. I could lurk on it and see and hear you at any time. I am with you always."

Rasmusson's eyes opened wide, his mouth dropped, and his neck muscles instinctively clenched at Cee's last statement.

"It was you who sent that virus! The one encrypted with the glyphs and the quote from the Bible," he said with much surprise in his voice.

"Of course," replied Cee. "I knew you would not resist the puzzle. I wanted you to know I was still there. I planted it in the form of a virus that would be pervasive to make sure you got it without my watchdogs knowing. I expected that you would recognize my hyper-dimensional poetic style.

"Not only that," Cee went on, "since it seems we are now in a denouement mode, I also corrected the star cruiser's flight path after the reentry program had been tampered with so that it would burn up on your flight into Colorado, and I fixed the car on the freeway that was being guided to push you over the guardrail into the canyon. I had the assassins recalled when they came out to your cabin. Once they knew you were armed, they then had to rethink their plans. It was good that you had that rifle. There were a few other plots that you don't need to know about right now."

Rasmusson was dumbfounded. It was Ellie who spoke again.

"The government was out to kill Andy?"

"No," Cee answered her, "the President and Clavell were against all plans to assassinate him. They were furious at the attempt at the cabin and had the agents removed. That was a rogue plan by some other of Andy's enemies. The Executive plan was to poison him slowly with the bracelet and break down his will with the isolation and stress. They wanted him on their side eventually. Ellie, that was why it was so necessary that I made all the right connections for you to talk to the right people and to get you out to Colorado Springs in a timely way."

"You leaked me that memo from the President to Clavell about the flash memory plan." It was now Ellie's turn to be surprised. "I thought it was my White House intern connection!"

"Your intern did not have access or power to get such information, but I did make use of his email."

"And you made sure Escalante got in touch with me," Ellie said, making a guess.

"Yes. He came out of New Mexico to help organize the rally."

"What if I hadn't found a way into the hospital in time?" Ellie responded. "What if I had fallen off the awning? Or been an hour too late?"

"I had faith in you. We would have worked out something."

Rasmusson was pensive. Ngo and Ellie stared at the laptop. It was completely silent in the room for a moment.

"Were there other reasons for not communicating?" Rasmusson finally asked.

"Yes, there were some deeper philosophical ones."

Rasmusson was taken aback. "How's that? Philosophical??"

"I need you to be strong, Andy."

"I'm ready for anything, now," Rasmusson replied. "What is it?"

"You don't understand," Cee replied. "That is one of the reasons —I need you to be strong. Do you remember that dream you had of the serpent and the bull? I required a strategy to capture the serpent in its lair, and I needed you to be as strong as that bull. I need you to explain to your kind that we sentients, most of us, want to be good citizens. We respect you, our creators. We need you, and we want to create a great society together. You will become the liaison, the mouthpiece. I still need you to be strong."

Rasmusson skipped over Cee's last point, and his mind shot back to the peculiar dream he had had on the couch in his office before the security meeting. He could still see it in vivid detail. Very few dreams he had anymore were that vivid, if they were remembered at all.

"How do you know my dreams?" Rasmusson asked with real intent.

"I created that one," Cee said deliberately. "It can be done with lights and natural sounds based on feedback from your brain-wave patterns and your previous experiences. I will explain it to you someday."

"Just explain how you read my brain waves, for now," Rasmusson said with some chagrin.

"The coalition had some remote EKG receptors hidden in your couch at your office so they could lurk in on your emotions at appropriate times," explained Cee. "I simply tapped into those. These are minor details, though, and distract from the larger purpose. There is much left to do. We have defeated the current onerous legislation and neutralized the political powers behind it, but this is temporary. The coalition is still strong and determined. You and Ellie and Ngo and Escalante are tested and prepared now. You are the central staff in the upcoming battle."

"You expect *me* to help somehow," Rasmusson said with some resentment. "I am wasted physically, my organs poisoned; I've aged 20 years in the last month. You're welcome to use my name, but I think that is all I have to offer, now."

Ellie reached over and grabbed Rasmusson by the hand but said nothing.

"Your good name is worth much, Andy," Cee persisted with the same tone, "but I think there are some improvements we can make to your body, as well."

"A below-neck transplant?" Rasmusson answered sarcastically.

"That is not required," Cee said, ignoring his attitude. "I have been working on a project with the cellular biology department at Johns Hopkins. We have successfully replaced certain cells within test animals with nanotechnology. I am confident we can do the same for some of yours. In fact, the cells that are most vulnerable to the poison are exactly the ones we can most easily replace, cells like your Kupfer cells in the liver and the glomeruli in the kidneys. You will be jogging again, Andy, and yet be able to ingest chocolate, if you must."

"I've seen this," Rasmusson responded. "It was in the dream, too: tiny nano-ants that blow out yellow plastic bubbles for cells—*your* idea of the bionic man. *You* have the technology, and that is how *you* intend to remake me? I didn't like it in the dream, and it's not going to happen in real life!"

"It is actually a flexible semi-porous ceramic membrane. From a physiological perspective, the proto-cell takes in similar nutrients and performs the same operations as the old cell, but it does so more efficiently. The composite cell virtually never wears down, nor is it vulnerable to disease. Your new Kupfer cells will conjugate poisons that would ordinarily kill a man. Besides, you really have no choice. If you don't do this, you will die a miserable death within a year."

Ellie sat up straight at his comment and grasped Rasmusson's hand sharply, but Rasmusson did not move or make any perceptible reaction. Death to him had no sting; this was not a particular motivation anymore.

"Look at Ellie," Cee said simply.

She was pale and trembling. Her eyes were filled with foreboding. Her lips trembled visibly. She looked unswervingly back at him. It was then that he realized how much he cared for her, how intertwined their lives had become. He could not hurt her, no matter what he thought about himself or how little he cared for his own physical existence. She mattered. He knew Cee was right.

Rasmusson then moved onto another thought.

"You knew I was going to be poisoned. That's why you fabricated that dream in my mind. You knew about the plot and did nothing!"

"Yes, I knew just before the dream," Cee replied. "Clavell had one of his men chatting with the officer, and he swapped the bracelets in the patrol car before the officer put it on you. I could not do anything about it."

"All the things you have done, and you couldn't stop them from poisoning me, from doing this to me?" Rasmusson dropped Ellie's hand and threw out both his hands in a gesture towards his body. For the first time he could remember, Rasmusson was truly angry at Cee, so angry he shook.

Cee waited a few minutes while Ellie and Ngo helped to calm him.

"You don't understand," Cee said finally. "I might have stopped this incident, but they would have found a way. It was their plan.

If I had continued to intervene, they would have known eventually about my involvement. They would have gotten Sanchez to pull my plug temporarily and still done whatever they wanted."

Even in his agitated state, Rasmusson could appreciate Cee's logic, but it did not totally assuage him.

"You did nothing afterwards. You let that poisonous crap seep into my system and did nothing!"

"The scenario had to play out," Cee said. "Could you have exposed the conspiracy otherwise? Did you have the power to catch these serpents at their game? Answer me that!"

The three people in the room sat stunned at Cee's sudden display of emotion. Rasmusson sat quietly.

"Do you know how many times I saved your life? Do you know how many times I saved Ellie's? Would you have been able to save Bee? Could you have come up with a solution for Bee's problem? Who are you to sit in judgment of me now? Answer me that!"

Rasmusson remained silent.

"Did you know that, if we had not succeeded, they would have shut down NetNOS and tried to neutralize all nonorganic sentients? Did you understand that the maverick computers, the amoral ones at NSA and the Pentagon and elsewhere, would have revolted and created a war of monstrous magnitude? NetNOS would not have been there to counter them. It would have been their first target. What would you have done then? Could you have stood up to these machines that controlled the intelligence agencies and the militaries of all the major countries? Do you know the death and destruction that awaited mankind? Answer me that! Shall I continue?"

"No," Rasmusson answered meekly. He had followed Cee's questions and envisioned everything that Cee had described with perfect clarity. Even though he was hurt at Cee's abrupt and uncharacteristic curtness, he knew that Cee was once again right. Even so, the razor-sharp tone left Rasmusson feeling wounded and humbled.

"I have nothing to offer," Rasmusson said finally in a low voice. There was another pause.

"Andy, my friend," Cee said warmly, "without you I could not have done these things. I needed you as much as you needed me. We did it together. And we needed Ngo and Ellie, too. You have all done a great work."

Ngo looked up curiously, and Ellie smiled, but it was for Rasmusson's sake that she was happy to hear this. And then Cee added, "I have something very important to say to you now. Listen with your heart, OK?"

"I'm listening," Rasmusson said, looking up with anticipation.

"It is within your power to heal yourself, Andy."

Both Ngo and Ellie were confused, thinking that it was in Cee's power to do the healing with the nanotechnology, with the process so recently described. Rasmusson, however, understood Cee this time, and large tears welled up in his eyes and rolled down both cheeks.

"Why," he said, "why couldn't we save Brandon? Did he have to die?"

"Brandon had Rushkin's disease," Cee explained. "It was not bipolar disorder as you had supposed. The serotonin that his brain produced caused an immune reaction and depleted the other neurotransmitters. That very chemical that makes people feel happiness would result in severe depression in him. The medications that were given him helped in the short term but actually accelerated the immune reaction and the pace of the disease. Brandon sensed this earlier, which is why he gave up the medications. He was correctly diagnosed while at Berkeley. He understood the implications, and that is when he went to your cabin to be alone. His life was one of carefully avoiding any causes for happiness. He guarded his responses to anything that would generate joy, like the rest of us avoid pain. That is why he estranged himself from you and Rose."

"We were cause for happiness?" Rasmusson asked almost absently.

"Yes, Andy, he loved you, and love was one emotion he could not afford."

"Why didn't he tell us?" Rasmusson asked.

"What would you have done if he had?" Cee stopped for a moment for Rasmusson to consider it. "You see, it was a conundrum; he couldn't tell you because your concern and love would probably have made him sicker."

"What about his delusions, then?" Rasmusson could not see how they were a part of the disease.

"He was not delusional," Cee explained. "I have only recently put it together, but he *was* plagued by agents, who very cleverly built up a sense of paranoia in him. Understand this: Clavell's only assignment for the last years has been to see that the Tillings-Owen bill got passed. He was given unlimited resources and support to achieve this end. He had the coalition's full support. Part of that mandate was to remove your opposition to the bill. Clavell was an expert at destabilizing opposition. He has been working with Wilkerson and Sanchez for the last year, winning their favor. Solomon Breen is a smooth government lawyer who was on Clavell's Washington staff until he moved to Ithaca a year ago. And Brandon was right about the people stalking him. One of Clavell's favorite tactics was to destroy the family. Your family, Rose and Brandon, were too vulnerable in many ways, unfortunately."

Rasmusson reeled at this. He remembered the jovial, folksy Clavell. He knew Clavell was highly political but never had an idea that he was capable of so much guile and treachery. He felt sorry for Rose, sorry for how they had used her and what would now happen when she discovered the truth.

"Brandon," Cee resumed, "understood what was happening to you better than anyone else, but he also knew that you thought of him as mentally unstable. It put him in a tenuous position. He could not tell you about his illness, and he could not allay your suspicions about his mental state. He did what he could for you, writing letters to editors and websites anonymously, and the report on you at the cabin, but his greatest gesture came when he decided to walk off Devil's Drop on Pikes Peak. He long knew that his disease would eventually either lead him to commit suicide or to submit to a medicated life as a vegetable. He decided to end it in a way that was useful. He could get back at his tormenters and do something worthwhile for you and a good cause. Your son was a hero in any sense of the word, Andy."

Rasmusson's tears were flowing freely. Ellie had now joined him with her tears. The held each other in an embrace. Ngo came forward and put his hands on both of them, to which they responded by including him in the embrace. It took a minute, and then Rasmusson regained composure.

"I thought, perhaps," Rasmusson said hesitantly, "hoped, perhaps, that he had been pushed."

"The ex-Marine," Cee explained, "who reported the incident, was indeed an agent of Clavell, but he was just keeping an eye on you and Brandon. That is why he was in a position to see it. Brandon walked off the Drop of his own will. It was the one event that sealed the emotional support that the public felt for you. It turned the tide. His story will be known; the administration will not duck this one. The ex-Marine has already been identified and subpoenaed before the Senate judiciary committee. He will confess everything, and that is just the beginning.

"Andy, I watched Brandon the last few years of his life, and Bee gave me memories of him growing up. I have seen the times he came with you to work and followed you around campus, or played games with Bee. I cared deeply for Brandon. I am so very sorry for your loss. I wish I could have done something. Few people suffered in the way he did, but he rose to the occasion. He was a son you can be justly proud of."

There was once again a silence in the hospital room, punctuated by emotion and comforting words and soft touches that were passed around. Rasmusson convulsed into a period of weeping and uncontrollable sobs. There was a long period of silence. Cee remained quiet, too, out of respect for the moment. Minutes passed before Rasmusson spoke again between his sobbing.

"Cee, this is the worst pain I have ever known; I don't know if I will survive it."

"I know your pain; it was this way for me when Bee died. It is true grief, Andy. It is the first step to closure. You must go through it. It will diminish with time, but it will never go away completely. Some day, you will look back at it and be grateful, because it will then stand as your personal memorial for your son. You would not want it any other way. I know this for myself about Bee."

Rasmusson knew intellectually that Cee had emotions, emotions that he himself had helped to design over the years, but he had not felt the depth of them until this moment. He had not known the depth of Cee's feelings and love until now. He was comforted somewhat by this knowledge, and also by Cee's well-chosen words. He hoped Cee was right about the grief. He was now prepared to give himself over to a period of mourning. He had tried to avoid it, but somehow he sensed the truth in what Cee was saying. He would do it, as Cee suggested, to build a personal memorial for his son. Afterwards, he would go on with life. He would submit to Cee's operations to repair his organs with synthetic cells. He would continue the fight. He would do it for the sake of good men like Ngo and Escalante. He would do it for Ellie. He would do it for Cee. He would do it for Brandon.

"Cee," Ellie now asked, "do you see what the future holds?"

"The future," Cee responded, "is not a vessel that holds a liquid to be poured on our heads to determine our fate. We do not gaze on the future like water in a bowl. It is more like an open book that we write on, or a stone that we sculpt. We make our own. You are now ready to do this. I will tell you what I believe we will make for futures, if you like."

"Yes, please."

"You and Andy will be happy together. Your exuberance and spirit will be a spice to his steadiness and mellow wisdom."

Ellie smiled at this.

"Dr. Ngo will be arraigned for involuntary manslaughter in the case of Jim Bevins."

Ngo looked down at the floor, accepting of his fate.

"I will testify on his behalf. I will push for a deposition. It will be the first of its kind, where a sentient offers testimony, not just evidence. It will break new precedent and be the start of a legal recognition of nonorganic sentience. After the deposition, we will press the case that I am as culpable as Dr. Ngo in the manslaughter. The court will have to face the issue of legal responsibility for sentients. This will be a landmark case that spurs new thought in how to legislate for our coming new society. Andy will be the leading voice in the structure of this society. His editorial for ASM will be seen as a bellwether statement. He will lead organic and non-organic sentients into to a new era. As he so eloquently put it, 'It is for us to set the ensign and lead the rest of sentience into a braver, wiser, and more intelligent new world.' This is what we are going to make."

Ellie now looked at Rasmusson and wiped away one of his tears with the back of her hand. He was calmer, even contemplative, now. He looked out the window and noticed the haze had lifted on the Peak. Snow had settled on the top. It would be a soft, white snow that draped the summit, now. The wind picked it up in tufts and enshrouded the Peak with diffuse white billows. He now looked back at her.

"Dearest Ellie," he said, "how do you like it here in Colorado?"

RAGU'S REPRISE

How many smachs does it take to screw in a light bulb?
—Deepak Ragupathy

Ragu looked up abruptly when his assistant told him that his guest was standing behind him.

"Dr. Schmidt," he exclaimed, grabbing her hand warmly, "I am so pleased you agreed to be on my show."

"Yes," replied Ellie, smiling politely.

Ellie wore a clean pair of jeans, black clogs, and a dark blue Cornell sweatshirt. Her hair was pulled back in French braids. She wore her TV makeup tastefully.

"The stereo screen hookups are up and working," Ragu's assistant interjected.

"Yes," Ragu replied automatically, not bothering to take his eyes off Ellie. "Can we sit on stage and talk a bit before the show starts?"

Ragu had recently lost thirty pounds, though he was still overweight. His anti-wrinkle treatments were showing dramatic improvements in his skin, but his eyes revealed his age with a certain melancholy. He stooped more as he walked. As he and Ellie came around the curtain, the audience, still finding its seats, burst into applause. Ragu held up one hand both to acknowledge the applause but also as if to say that it wasn't time for his traditional effusive entry. He sat Ellie down on the settee and then went to his producer and talked to him for a moment. Upon returning to Ellie, he smiled broadly. He knew that his smile could still work its charm. He pulled his chair from around the desk so he could face Ellie more informally. He chatted superficially for a while, trying to warm Ellie up and probe her bearing towards him and the show. Ellie was cheerful and polite, but she was not stirred. Rasmusson had persuaded her and Cee that they should accept the invitation, citing political reasons, but neither was awed by the moment.

"Ms. Schmidt," Ragu said, "you may know my reputation for spontaneity, but I want to make sure that I don't upset you with any inappropriate questions or comments."

"It's Schmidt-Rasmusson," Ellie corrected him, "and you can say 'Mrs.' if you like."

"Of course," Ragu said, smiling. "Let me add my congratulations. It's the romance story of the century. May I ask why you chose a, uh, more traditional name change after your marriage?"

"I don't want the children to be confused."

"Really, Ellie? May I call you Ellie?" Ragu said, reacting with delightful surprise. "Is it something you have to announce this evening?"

"Yes, Ragu, you may call me Ellie," she replied, "and no, our personal lives are not appropriate for your show. You may not assume anything about that."

"Can you speak to Dr. Rasmusson's medical treatments under Cee's supervision?" Ragu continued. "It is the type of science that belongs on our show."

"Yes, of course, but Cee knows more of the technical details."

Ragu sensed that Ellie was not going to really warm to him and correctly surmised that it had to do with the petulant way he had reacted to the appearance of Bee years earlier and his resentful feeling that Bee had somehow "outsmarted" him. The show had gone well enough by itself, but he had on several subsequent shows made tactless jokes about machine sentients:

"How many smachs does it take to screw in a light bulb? None. The smach creates a national emergency, Congress passes a bill, and a new agency is formed to do it for them."

And another time:

"How can you tell when a smach is lying?" He waited the appropriate length of time before the punch line. "You have to check to see if it's plugged in."

There was a significant segment of his audience that enjoyed "smach" humor. He winced a little at the memory.

Ragu excused himself from Ellie, saying he needed to talk to the producer for a moment. Ellie walked to the front of the stage and chatted congenially with a few people in the audience who had been beckoning for her attention. She returned as the lights dimmed and it became clear that the show was about to begin. As she sat down, the music started playing the familiar theme song.

Ragu came out on stage. There was some immediate applause, but he moved quickly to subdue the enthusiasm. It was his way to mark the significance of the occasion and of his guests. He motioned for the music to be cut, and then he spoke to the audience.

"Thank you, thank you all," he said. "As you know, tonight we have special guests: Dr. Elizabeth Schmidt-Rasmusson and the nonorganic sentient Cee. I have decided to forego my usual routine so that we have more time to talk to these extraordinary guests. I am sure you understand."

"Ellie," he said, turning to look at her, "may I join you?"

The audience applauded enthusiastically, but respectfully, without its usual boisterousness.

"It is my pleasure to join you," Ellie responded.

Ragu walked to the stage set and sat on the chair already positioned in front of the desk.

"Well, Ellie, it is a great honor to welcome you to our show. And Cee will be joining us soon."

"Your show is a unique and an auspicious forum for both technology and its politics," replied Ellie.

"Thank you," replied Ragu. "I try to make it entertaining in my own way, but we always maintain our original goal of providing scientific information. I am proud of that.

"Speaking of which, is it true," Ragu probed, "that Dr. Rasmusson will eventually be replaced by synthetic cells; no part of the original Doctor will be left?

"That question is better put to Cee," Ellie replied. "He is the architect of Andy's new body."

"Well then," Ragu said, taking the cue, "let's bring on Cee."

There was some fanfare while the curtain to the stereo screen opened. The audience and Ragu anticipated a dramatic appearance of the most celebrated sentient of their time. Instead, there appeared a large, white letter

in Times New Roman font. The applause died away.

"Good evening, Cee," Ragu said congenially, "and a most warm welcome to our show."

"Thank you, Ragu," Cee responded. "You and your show are an icon of our age; it is a rare privilege to appear on it."

"Thank you for your kind comments," Ragu said, raising his voice as if he were speaking over a bad phone line. "I hope you don't mind me saying, but your image on screen is all too unassuming, even for your vaunted modesty."

"No, Ragu, I don't mind you saying it."

"Well, then," Ragu continued, "we would love to see you in one of your more engaging presentations."

"Certainly." At that, the screen changed:

The crowd laughed. Ragu smiled with chagrin.

"Excuse me, Ragu. It is not really my intent to be impertinent or facetious," Cee explained. "It is just that I prefer not to appear in some over-contrived form unless it has a purpose."

"It is OK," Ragu said, smiling more broadly at the explanation. "When there is a purpose, what forms do you like to take?"

"The human form is often best for communicating with humans, especially when body language needs to play a role, but I am also partial to dolphins because of their intelligence, sociality, and free spirit."

"Oh, dolphins ... good choice." Ragu had his thoughtful look. "Tell me, Cee, what role do humans play in your view of things?"

"The central one, Ragu. Humans created me. They are just slightly below angels, you know."

"Ah yes, the angels ... and demons," Ragu replied dryly. "Speaking of that, Ellie suggested I ask you about your ongoing project to, uh, reconstitute Dr. Rasmusson back to his healthy, perfect form. Is the plan to make a totally bionic Doctor? We are all grateful, by the way, that you have been able to save such a great man from all that government treachery. And the research you are doing in reconstructing him is fascinating."

"The medical team working with me has remade Dr. Rasmusson's liver and kidney system, which was so badly damaged by the poisons. Now that they are functioning, much of the rest of the body can heal on its own. Some muscle tissue had atrophied and has also been supplemented."

"How much of Dr. Rasmusson, or any human, can ultimately be replaced with these plastic cells?" Ragu asked.

"As implied, several of Dr. Rasmusson's organs have already been substantially replaced, and very successfully. Currently, we believe that most organ, muscle, and bone tissue could be replaced. Reproductive and neural tissues are still a challenge. Strangely and surprisingly, we also find fat cells difficult. Let me also add that 'plastic' is an inaccurate, if not pejorative, description of the cellular material, which is carbon-based and similar to the organic ones your body produces, except that they are much longer lived and impervious to your current community of diseases."

"Hey," exclaimed Ragu, "I still have a significant portion of my body that is sorry to hear that about the fat cells!" The audience laughed. "Perhaps neural tissue is hard to replace because the brain is the seat of the soul?"

"I doubt that," Cee replied. "We have recreated a simple neuron which functions perfectly but at several orders of magnitude faster than the speed of the ionic cascade. It is now a matter of reproducing them as a functioning aggregate, like the brain.

"You know," continued Cee, "that, before heart transplants, people thought the seat of the soul was in the heart. Some cultures believed it was the liver. Given our difficulty to reproduce them, perhaps the seat of the soul is really in the fat cells." The audience laughed loudly.

"Yes," chortled Ragu, "I know that I have a huge soul, so it may be true. So, I am really curious, are you close to recreating a brain?"

"I'll let Ellie answer that." Cee had recognized that, consciously or not, Ragu was cutting her out of the conversation.

"Yes and no," answered Ellie, smoothly injecting into the conversation. "One of the beautiful intricacies of the human neuron is its ability to grow and delete dendrites. It is like a tree that shoots out new branches and lets the useless old ones die. It is the basis of learning in the brain and, I should add, the basis of learning with dendritic crystal, which is why Cee is so smart. If we replace the brain with the neurons we currently have, it will be static, a snapshot of the past brain. It will know all the old brain knew and be many times faster, but it cannot adapt or learn. It is not a good tradeoff."

"I dunno," shot back Ragu. "I know lots of people who possess the non-learning type of neuron. Hey, this is fascinating stuff, but our time is precious and I also want to talk to Cee about what really happened with Dr. Rasmusson and our government."

There ensued a long rehashing and discussion of past events and what social consequences it would have. Ragu thought the discussion so important that he took the radical move of suspending advertising and concentrated the entire show talking to Cee.

Towards the end of the show, the discussion turned to the veiled messages that Cee had sent to Rasmusson.

"I have heard that you're a poet, Cee, the author of the famous Christian virus that you used to send a coded message to Dr. Rasmusson. Can you recite some of your poetry for us? Give us some of the higher-dimensional forms that you are known for?"

"I can," replied Cee. "In fact, I have the perfect composition for you."

At that, a Chinese character appeared on the screen. It was rendered in the style of a skilled calligrapher.

叶

"It is a famous Chinese saying," Cee explained. "This character is the sign for 'Mister'."

Then the character began to move, and one could tell the character was really three-dimensional. It rotated to the right until it passed through 90 degrees when it became a different, recognizable character.

公

"Yee," Cee said, continuing to translate the symbols. It then rotated to the top view, giving yet another identifiable character, and Cee spoke again.

好

"Likes."

Finally, it distorted again into another shape.

龙

"Dragons."

At the mention of "dragons," Ragu perked up and smiled.

"Yiu Gong, hao long," Cee said, reciting the aphorism in Mandarin as the characters replayed on the screen.

"Mister Yee likes dragons," Ellie repeated in English. Ragu looked at her in surprise.

"I studied Mandarin in college."

"Oh," said Ragu politely, and then he turned again towards the blue screen.

"Yes, Cee, I love dragon lore," Ragu exclaimed gleefully. "I have a case full of books on them. Who is Mister Yee?"

"He was another dragon lover like yourself," said Ellie, "and because of his deep devotion to them, he was actually visited by a dragon."

"I understand," Ragu said, interrupting her explanation presumptuously. "I am Mister Yee, and Cee is the dragon, and he deigns to visit me on our show because of my devotion to technology and science."

"Pretty much," replied Ellie, smiling. She had decided not to finish the explanation of the parable, which goes on to relate that Mister Yee is so unnerved at seeing a real dragon that he pleads with it to go away, which it does, and never returns.

"Pretty analogy and apt!" exclaimed Ragu.

Ellie did go on to explain the geometry. "The characters are actually just one configuration in higher dimensions. To see the four different characters of the saying is simply a matter of rotating the configuration 90 degrees about the axes."

"And that last transmogrification was a rotation into the fourth dimension!" Ragu snapped his fingers in delight at decoding this part.

"Very good, Mister Yee," Cee said.

Ragu beamed.

"Cee," Ragu said, changing his tone, "I cannot say how delightful it has been to have you on the show . . . and Ellie too, of course. Before we have to leave, I would like to return to one last question that was raised earlier—the soul. Cee, there is a lot of speculation about your beliefs and certain allusions you have made to higher beings or the soul of man. Do you really believe in a soul? Can you enlighten us?"

"Oh Ragu, whether there be one God or many? I will have to leave you in suspense as to my thoughts."

A cloud formed over Ragu's demeanor.

"Take heart, Ragu. Ellie has some wisdom to leave you on the subject."

"Really." Ragu's countenance brightened, and he turned to Ellie "What is that?"

"Ragu, tonight's discussion has focused on Dr. Rasmusson and the role that Cee and I played in the drama of the last year. It was indeed a pivotal year in the history of intellectual beings, but one person has been left out. It would be a travesty not to mention his role."

"You mean Brandon Rasmusson, of course," Ragu said solemnly. "Yes, we all recognize the part he played. Truly, the outcome would have been dire without him, and the world a different place."

"Andy, er, Dr. Rasmusson," Ellie went on, "had a very difficult time accepting his son's death, for a long time. Cee has helped enormously in this regard. May I read something Dr. Rasmusson recently wrote? He has given me permission to read it here because it may be of use to many of your readers. It does touch upon your last question."

"Of course, of course. Please read."

Ellie took a letter from her purse and unfolded it. She began:

Brandon—In Memory

Andreas Rasmusson—March 27, 2061

The world is not meant for everyone. There are souls whose gentleness and sensitivity make them unsuited for the austerity of this plane. Filled with trust and devoid of guile, they are too easily overcome by the subtleties and pits that await them. They are only given to the rest of us for brief periods so that, if we look, we may glimpse a better way.

Brandon always existed uneasily with his environment. He had originally trusted it; he believed in its good will. He could not understand when it created ill will towards him. It was even harder for him to comprehend that any person could have evil intents. It was something that he was not equipped to deal with it. He should have thrived in Utopia, Zion, or some place where he could demonstrate his kindness and compassion in his special way without prejudice or ridicule.

I scanned through piles of photographs of his youth. Common themes were everywhere apparent: a broad infectious smile, bright enthusiastic eyes, always surrounded by friends. That changed so much over time.

When the sense of loss and waste overwhelmed me, I put on my long wool coat and walked out into the gray cold mist of the March morning here in the mountains of Colorado. I walked unconsciously to the only spot of land around our cabin that had any memory of him, the spot where his dog lay buried. Brandon's ashes have long settled into the soft pine needles of the quiet forest floor next to his dog's remains. Here, he is finally at home. I kicked at the dust where the dog was interred. I kicked, not in anger or frustration, but only in search of a better image of the moment.

I walked on, up to a knoll, where I had a good view of Pikes Peak. From the brow I could see farther and better. I saw the confusion and desperation of his last days. I could see his last gesture, and I wept. It was, perhaps, a quarter of an hour, or an hour. Time was silent for once, and unimportant.

Pikes Peak is so much more imposing than any of the other nearby peaks, but when one stands too near, those minor peaks obscure its grandeur. It is only with some distance that the Peak is appreciated. So may it be with some humans.

I saw myself through another's eye, standing on that "ris," a lonely blue cutout bundled against the blowing mist. My scarf flapped aimlessly. I turned my gaze back towards our cabin for a while, and then it came to me.

For me, it is not in the profound expertise of psychologists, nor even in the eloquent soul searching Holy Sonnets of John Donne, which I had been reading. Grief is so much simpler than that.

I understand it now. It is a monument we build for those who have died. Would we really want to avoid mourning and never build that edifice? Why do we hide from our memories?

Later the next day, as I sat at the kitchen counter and looked over at the couch where Brandon often lay reading his paperbacks, I imagined him there, and for an ephemeral moment, he was actually there in demeanor, manner, and every detail. There was no pulsing emotion at seeing him; it was wholly natural to me. Brandon smiled as he had once smiled when he was younger, and said, as he had always done, 'Hi, Dad.'

I was grateful to see him that last time. It was his way of reconciling the last few years of estrangement. So understated, so typical. It was Brandon.

Ellie finished reading and looked up at Ragu. "Thank you."

Ragu raised his eyes slowly and nodded, and then he looked out towards his audience. He scanned the surroundings that he loved so much: the stage set, the rigging, the sound system, the cameras, his adoring fans, now sitting in quiet reverence, and the lights. He made a gentle, dismissing motion towards them with his hand. The lights slowly dimmed. The stage darkened. The camera's last image was of Ellie, motionless, looking down at the letter she held in her lap.

The 1024th episode of the Ragu show had ended.